Landscape of Next Generation Sequencing Using Pattern Recognition

Performance Analysis and Applications

RIVER PUBLISHERS SERIES IN BIOTECHNOLOGY AND MEDICAL RESEARCH

Aiming primarily at providing detailed snapshots of critical issues in biotechnology and medicine that are reaching a tipping point in financial investment or industrial deployment, the scope of the series encompasses various specialty areas including pharmaceutical sciences and healthcare, industrial biotechnology, and biomaterials. Areas of primary interest comprise immunology, virology, microbiology, molecular biology, stem cells, hematopoiesis, oncology, regenerative medicine, biologics, polymer science, formulation and drug delivery, renewable chemicals, manufacturing, and biorefineries.

Each volume presents comprehensive review and opinion articles covering all fundamental aspect of the focus topic. The editors/authors of each volume are experts in their respective fields and publications are peer-reviewed.

For a list of other books in this series, visit www.riverpublishers.com

Landscape of Next Generation Sequencing Using Pattern Recognition

Performance Analysis and Applications

Saurav Mallik

The University of Arizona, USA

Loveleen Gaur

Amity University, India

Soumita Seth

Future Institute of Engineering and Management, Kolkata, India

Tapas Bhadra

Aliah University, Kolkata, India

Mingqiang Wang

Stanford Cardiovascular Institute, School of Medicine, Stanford University, USA

River Publishers

Routledge
Taylor & Francis Group
NEW YORK AND LONDON

Published 2024 by River Publishers
River Publishers
Alsbjergvej 10, 9260 Gistrup, Denmark
www.riverpublishers.com

Distributed exclusively by Routledge
605 Third Avenue, New York, NY 10017, USA
4 Park Square, Milton Park, Abingdon, Oxon OX14 4RN

Landscape of Next Generation Sequencing Using Pattern Recognition / by Saurav Mallik, Loveleen Gaur, Soumita Seth, Tapas Bhadra, Mingqiang Wang.

Routledge is an imprint of the Taylor & Francis Group, an informa business

ISBN 978-87-7004-151-5 (hardback)
ISBN 978-87-7004-205-5 (paperback)
ISBN 978-87-7004-641-1 (online)
ISBN 978-87-7004-209-3 (master ebook)

Contents

Preface

This book focuses on Next Generation Sequencing (NGS), a groundbreaking technology that has revolutionized the examination of organisms and significantly impacted biomedical research and disease diagnosis. The rapid advancement of large-scale NGS technologies has introduced numerous computational challenges, including metagenomic classification, gene module identification, sub-graph clustering, gene signature detection, multi-omics data integration, spatial transcriptomics, and other downstream analyses for big biomedical sequence data.

Biomedical imaging involves the use of various techniques such as X-rays, CT scans, MRIs, and ultrasounds to obtain images of internal organs for diagnostic, treatment planning, follow-up, and surgical purposes. Deep Learning (DL) algorithms have garnered considerable interest for their applications in both NGS and imaging, particularly when these data are used for the same group of patients.

In biomedical research, disease classification, feature dimension reduction, and heterogeneity are the primary research phenomena. Clinical researchers utilize AI approaches to efficiently analyze complex biomedical datasets, such as multi-omics datasets. Single-cell RNA-sequencing (scRNA-seq) enables the examination of gene expression levels in individual cells within a heterogeneous population, providing significant sample sizes to leverage robust AI technologies.

As single-cell sequencing techniques become more affordable and data more accessible, researchers require the next generation of AI tools tailored for scRNA-seq data. By leveraging NGS data and biomedical imaging, researchers can utilize deep-learning models to predict diseases, accelerate disease detection and diagnosis, reduce treatment expenses, and enhance public health. This book covers a wide range of critical disease data and medical images.

RStudio is a versatile open-source integrated development environment (IDE) widely used as a graphical interface for R version 3.0.1 or above. Additionally, it supports various other programming languages, including

Python and SQL. The official R website is The Comprehensive R Archive Network (CRAN), hosted by the Institute for Statistics and Mathematics of Wirtschaftsuniversitat Wien. For product information, please contact the Institute for Statistics and Mathematics (https://www.wu.ac.at/en/statmath/).

List of Figures

List of Tables

List of Abbreviations

AE	Autoencoder
AI	Artificial intelligence
ARM	Association rule mining
AT	*Arabidopsis thaliana*
AUC	Area under curve
BH	Benjamini–Hochberg
BP	Biological process
CAM	Cell adhesion molecule
CART	Classification and regression tree
CAT	Computed axial tomography
CC	Cellular component
CDR	Cellular detection rate
CDSS	Clinical decision support systems
CE	Capillary electrophoresis
ChIP-seq	Chromatin immunoprecipitation sequencing
CNN	Convolution neural network
CNV	Copy number variations
CT	Computed tomography
DEG	Differentially expressed gene
DMR	Differentially methylated region
DNA	Deoxyribonucleic acid
DNMT	DNA methyltransferase
DOM	*Drosophila melanogaster*
DR	*Danio rerio*
DT	Decision tree
DWT	Discrete wavelet transform
EBPSO	Enhanced binary particle swarm optimization
FD	Fractal dimension
FDR	False discovery rate
GA	Genetic algorithm
GC	Guanine-cytosine

GDC	Genomic Data Commons
GEO	Gene Expression Omnibus
GESA	Gene set enrichment analysis
GO	Gene ontology
GRB	Gaussian radial basis function
GSTR	Geneset-tree
HE	Hurst exponent
HPRD	Human Protein Resource Database
HS	*Homo sapiens*
HSTS	High-grade soft tissue sarcomas
ICA	Independent component analysis
ID3	Iterative Dichotomiser 3
IG	Information gain
IHPS	Isolated hybrid power system
KNN	*K*-nearest neighbor
KSVM	Kernel Support Vector Machine
LDA	Linear discriminant analysis
LIME	Local interpretable model-agnostic explanation
LSTM	Long short-term memory
MADGAN	Medical anomaly detection generative adversarial network
MAST	Model-based analysis of single-cell transcriptomics
MCDM	Multi-criteria decision-making
MCI	Mild cognitive impairment
MF	Molecular function
ML	Machine learning
MM	*Mus musculus*
MOOV ARM	Multi-objective optimized variable cutoff-based association rule mining
MRI	Magnetic resonance imaging
MSE	Modified Shannon entropy
NCI	National Cancer Institute
NGS	Next-generation sequencing
NIS	Negative ideal solution
nrTCL	Normal tumor cell line
OCT	Optical coherence tomography
PC	Principal component
PCA	Principal component analysis

PCR	Polymerase Chain Reaction
PET	Positron emission tomography
PIS	Positive ideal solution
PPI	protein–protein interaction
QC	Quality control
R/S analysis	Rescaled range analysis
ReLU	Rectified linear unit
RF	Random forest
ROI	Region of interest
RRBS	Reduced representation bisulfite sequencing
scRNAseq	Single-cell RNA sequencing
SE	Shannon entropy
SGLD	Spatial gray-level dependency matrix
SHAP	Shapley additive explanation
SMOTE	Synthetic minority oversampling technique
SNNS	Stuttgart Neural Network Simulator
SNV	single nucleotide variation
SOP	Standard operation procedure
SVM	Support vector machine
TCGA	The Cancer Genome Atlas Project
TF	Transcription factor
t-SNE	t-Distributed stochastic neighbor embedding
UdPL	Undifferentiated pleomorphic liposarcoma
UMAP	Uniform manifold approximation and projection
vst	Variance stabilizing transformation
WES	Whole-exome sequencing
WGBS	Whole-genome bisulfite sequencing
WGS	Whole-genome sequencing
WHO	World Health Organization
WTG	Wind turbine generator
WXS	Whole exome sequence
XAI	Explainable AI

1

Introduction: Fundamentals of Next-Generation Sequencing, Pattern Recognition, and Biomedical Images

1.1 What is Next-Generation Sequencing (NGS)?

Next-generation sequencing (NGS) is a technology that identifies the DNA or RNA sequence in order to explore genetic variation linked to diseases or other biological phenomena. This technique, which became initially available for commercial use in 2005, was initially referred to as "massively parallel sequencing" because it enabled many DNA strands to be sequenced simultaneously rather than one at a time as with conventional Sanger sequencing by capillary electrophoresis (CE). In the context of genetic analysis today, each of these tools is useful. Sanger sequencing can be completed in a single day and is excellent for examining small amounts of gene targets and samples. Sanger sequencing is also regarded as the industry-standard sequencing technology; therefore, NGS results are usually validated using it. In addition to allowing for simultaneous investigation of hundreds to thousands of genes across many samples, NGS also allows for the identification and study of a variety of genomic features in a single sequencing run, including single nucleotide variations (SNVs), copy number and structural variants, and even RNA fusions. NGS offers optimal throughput per run, enabling speedy and economical study execution. Additional benefits of NGS over Sanger sequencing include the ability to detect mutations at lower allele frequencies, higher precision, and lesser sample input requirements. Genetic analysis has been transformed by the speed, throughput, and accuracy of NGS, which has also opened up new applications in forensic science, reproductive health, and environmental, agricultural, and clinical research [1].

1.2 The Scope of NGS Technology

NGS can be executed at many levels. The whole genome can be sequenced with it. Nearly every nucleotide in the genome, including chromosomal DNA and mitochondrial DNA, is analyzed at this level. Whole genome sequencing is utilized less frequently in medical applications and more frequently in research. In healthcare environments, it is applied more frequently to constitutional genetic diseases than to somatic cancer mutations. For the diagnosis of several rare genetic diseases, it is extremely helpful. When a genetic disorder is suspected, for instance, but no specific mutation has been identified by other molecular tests, whole genome sequencing might provide more comprehensive information about the mutations associated with the disease. It is employed less commonly for cancer somatic mutation because the average depth of whole genome sequencing has limitations. Allelic mutation frequencies and the proportion of tumor cells in various specimens may change in a certain tumor. Deep sequencing is frequently required for the detection of various mutations with various allelic frequencies in such circumstances, which is particularly difficult for the whole genome sequencing method.

Sequencing a whole exome can be done using NGS technology. Any cell type inside an organism, including all of its axons, can have its entire coding area sequenced. That makes up roughly 1% of the human genome and is more frequently employed in research.

Additionally, NGS can be carried out at the transcriptome level, which entails an entire assembly of RNA transcripts in any specific cell type, including mRNA, rRNA, tRNA, micro-RNA, and non-coding RNA. This is referred to as RNA sequencing as opposed to DNA sequencing. To find fusion genes, specialized mRNA sequencing is often used.

The most popular NGS test for cancer patients is targeted panel sequencing, which often examines dozens or hundreds of targeted genes. Typically, these focused NGS assays are created for a specific disease or group of disorders, such as a panel created for myeloid leukemia or a panel created for carcinoma. These designed panels have fewer targets than whole genome sequencing. As a result, it enables much deeper sequencing, which is required to cover various mutations with various allelic mutation frequencies [2].

1.3 Phases of NGS Technology in Clinical Applications

1.3.1 Identify the target genes for NGS clinical application

This usually depends on which disease or diseases the specific NGS testing is intended to diagnose. By way of illustration, if a patient has cancer, the first thing that needs to be done is to look up the most recent recommendations for the disease or diseases in question to determine which gene mutations are specifically mentioned as the recommended approach of treatment. For instance, according to the guidelines, tests should be done for EGFR, KRAS, BRAF, MET, RET, and Her2 mutations as well as ALK and ROS1 translocations in non-small cell lung cancer. A lot of progress is being made daily in the field of cancer mutation research [2].

1.3.2 Identify the NGS technique for clinical use

The next step is to choose a suitable NGS technique for an NGS panel. Different NGS techniques are available. An amplicon assay can be utilized for a gene panel that is clearly specified. However, a different approach, such as a hybridization capture assay, can be utilized if some targeted genes make it difficult for an amplicon assay to function. For instance, CEBPA and FLT3 mutations should be included in a panel that has been designed for AML patients. The high GC concentration of the CEBPA gene, which presents a barrier for an amplicon assay, is well documented. Internal tandem duplicates of various sizes usually create FLT3 mutations [2]. Wet lab procedures and dry lab (bioinformatics pipeline) operations should both be included in the standard operation procedure (SOP).

1.3.3 Validation of NGS Test for Clinical Application

The validation involves a variety of tasks, such as ensuring assay precision, reportable range, reference range, analytical sensitivity, and specificity. It is necessary to outline a validation strategy. Typically, the validation plan specifies which samples will be employed as positive and negative controls. Positive controls with various allelic mutation frequencies can also be created by combining the positive and negative samples at different ratios. To evaluate the assay sensitivity, such controls will be employed. For example, single nucleotide alterations, deletions, and insertions should all be included in the positive controls. It is important to take into account the sizes of the deletions and insertions. Typically, the larger the deletion and insertion, the more challenging the NGS test will be to interrogate. The number of samples that are required to be tested needs to be defined as well in the validation plan [2].

1.4 Next-Generation Sequencing-based Omics

Integrating data from several omics approaches, including genomes, transcriptomics, epigenomics, and proteomics, is necessary to comprehend complicated human diseases. Here, we provide a brief summary of a multi-omics technologies used with the NGS platform [3].

1.4.1 Genomics

NGS-based genomic studies deeply analyze DNA applying a variety of techniques, including whole-genome, whole-exome, and targeted sequencing.

1.4.1.1 Whole-genome sequencing

The process of whole-genome sequencing (WGS), which entails figuring out the whole DNA sequence of an individual's genome, is robust and thorough. It gives a thorough picture of an individual's genetic make-up, including every gene, regulatory region, and non-coding element found in their genome. It is mostly used in discovery science fields such as population genetics, eukaryotic and prokaryotic novel genome assembly, cancer research, rare genetic diseases, patients with complex disease symptoms, and plant and animal research [4].

1.4.1.2 Whole-exome sequencing

The goal of whole-exome sequencing (WES), a sequencing strategy, is to accumulate and sequence the exome, or protein-coding portions of the genome. Though only making about 1%–2% of the overall genome, the exome is related to the majority of known disease-causing mutations. WES allows for the detection of genetic alterations, including as insertions, deletions, and copy number variations (CNVs), within protein-coding genes by sequencing the exome. When discovering variations for population and cancer genetics as well as rare diseases with clusters of symptoms, WES is a more affordable option than WGS [5, 6, 7, 8].

1.4.1.3 Targeted sequencing

Targeted sequencing can detect a variety of genetic variations, including SNVs, small gene deletions, duplications, insertions, or gene rearrangements associated with disease phenotypes. But compared to WGS or WES, it has less exploratory power. However, benefits include affordableness and manageable data for health professionals, making clinical decisions simpler with more accurate disease-relevant information. For rare alleles in genetic diseases, as well as for low-abundant evolving mutant clones born as a result of tumor heterogeneity or disease evolution in cancer, it can provide significantly deeper coverage of up to 5000 [9].

1.4.2 Transcriptomics

Transcriptomics has undergone an enormous transformation as a result of sequencing (NGS), which has revolutionized our abilities to study the transcriptome – the entire set of RNA molecules in an organism or specific cell class. Researchers may acquire profound insights into gene expression, alternative splicing, non-coding RNA control, multiple biological processes, and diseases due to NGS technologies offering high-throughput and cost-efficient methods for profiling and analyzing RNA molecules [10, 11, 12, 13]. The following are some crucial functions of NGS in transcriptomics.

1.4.2.1 mRNA Sequencing (RNA-seq)

A popular NGS application in transcriptomics is RNA-seq. It entails the sequencing and quantification of mRNA molecules to provide an entire overview of the expressed genes in a biological sample. NGS makes it possible for researchers to accurately identify and measure gene expression levels by generating millions of short sequencing reads. The analysis of RNA-seq data enables the identification of novel transcripts, evaluation of alternative splicing processes, and the study of the gene expression dynamics across time or across various tissues or cell types [14, 15].

1.4.2.2 Alternative splicing analysis

The process of alternative splicing, which allows a single gene to produce numerous mRNA isoforms, significantly increases to the complexity of the transcriptome. NGS makes it possible to thoroughly examine alternative splicing patterns. Researchers may detect alternative splicing events and identify splice junctions by matching RNA-seq readings to the reference genome. The measurement and characterization of transcript isoforms is made feasible by this information, which also sheds light on isoform diversity, tissue-specific expression, and the functional effects of alternative splicing [16].

1.4.2.3 Long non-coding RNA (lncRNA) and small-RNA analysis

NGS makes it easier to study non-coding RNAs, which are important for gene regulation. The identification and characterization of different categories of non-coding RNAs are made possible by methods like small-RNA sequencing and long non-coding RNA sequencing. Small regulatory RNAs, such as microRNAs, piRNAs, and snoRNAs, can be profiled by small-RNA sequencing, revealing information on their functions in post-transcriptional gene regulation. The discovery and study of long non-coding RNA transcripts, which have been correlated with several kinds of biological processes and diseases, became accessible by long non-coding RNA sequencing [17, 18, 19].

1.4.2.4 Transcriptome assembly and annotation

The transcriptome of an organism may be reconstructed and annotated using NGS data. Researchers can find novel transcripts, splice variants, untranslated regions, and other transcript properties by aligning RNA-seq readings to a reference genome or by utilizing de novo assembly techniques. With the use of this information, we may better comprehend the complexity of the transcriptome and annotate reference genomes, leading to the identification of previously undiscovered genes and regulatory elements [20].

1.4.2.5 Single-cell transcriptomics

Single-cell transcriptomics, which allows the analysis of gene expression profiles at the level of an individual cell, has developed as a result of NGS. Single-cell RNA-seq (scRNA-Seq) technologies enable the profiling of transcriptomes from individual cells, revealing information on cellular heterogeneity, cell type identification, cell lineage analyses, and dynamics of gene expression in complex tissues or developmental processes [21, 22].

1.4.2.6 Integrative transcriptomics

To acquire a thorough knowledge of gene regulation and biological processes, NGS data from transcriptomics may be integrated with other omics data from genomics, epigenomics, and proteomics. Integrative techniques enable the identification of significant regulatory mechanisms promoting cell functions and diseases and give a system-level depiction of molecular interactions [21].

1.4.3 Epigenomics

The study of epigenetic modifications, which are heritable variations in gene expression patterns without alterations to the DNA sequence, is referred to as epigenomics. DNA methylation, histone modification, and RNA methylation (epi-transcriptome) are the three most frequently studied forms of epigenetic alterations [23], [24]. The position of nucleosomes, chromatin remodeling, and DNA accessibility are all affected by these chemical markers. Environmental elements including nutrition, pollutants, toxicants, and inflammation affect these alterations. Scientists have learned more about these epigenetic changes, particularly DNA methylation and hydroxymethylation, due to the information and data obtained through whole-genome sequencing in humans, animals, and plants. Researchers and physicians have been interested in epigenetic changes as they relate to complex diseases like behavioral disorders, memory, cancer, immunological, addiction, neurodegenerative, and psychiatric disorders [25]. NGS has been used to study epigenomics, as will be covered below.

1.4.3.1 DNA methylation profiling

DNA methylation is an essential epigenetic alteration that is essential for regulating gene expression and cellular functions. Single-nucleotide resolution genome-wide evaluations of DNA methylation patterns are made possible by NGS. Several methods leverage NGS to find methylated cytosines, including whole-genome bisulfite sequencing (WGBS) and reduced representation bisulfite sequencing (RRBS). With the assistance of these techniques, scientists may investigate the dynamics of DNA methylation, identify differentially methylated regions (DMRs) associated with certain diseases, and better understand how methylation affects gene expression [26].

1.4.3.2 Chromatin accessibility mapping (ATAC-seq/DNase-seq)

The genome-wide analysis of chromatin accessibility is made possible by NGS-based methods like DNase-seq and the assessment for transposase-accessible chromatin applying sequencing (ATAC-seq). These techniques reveal areas of the genome that DNA-binding proteins and transcription factors may access, giving information on enhancers, promoters, and regulatory elements of genes. Researchers can identify the functional components of the genome by integrating information on chromatin accessibility with information on other epigenetic alterations, gene expression, and transcription factor binding [27, 28].

1.4.3.3 Histone modification analysis

Essential epigenetic markers that regulate chromatin structure and gene expression include histone modifications including acetylation, methylation, and phosphorylation. Genome-wide profiling of histone modifications is made possible by chromatin immunoprecipitation sequencing (ChIP-seq), which pulls down the protein using an antibody before enriching the DNA that is bound to the protein and sequencing the results. This method is useful in a wide range of scientific fields, including the analysis of DNA histone modifications, the identification of transcription factor (TF) binding sites, and DNA methylation. Antibodies that target histone modifications are used to pull down DNA and sequence it using the NGS method in order to study these alterations. In order to identify the histone modification patterns at specific genomic regions, the resultant reads are aligned to the reference genome. ChIP-seq can provide insight into the chromatin states of cells, the discovery of enhancers and other regulatory elements, and the epigenetic control of gene expression [29, 30, 31, 32].

1.4.3.4 Chromatin conformation analysis

Hi-C and 4C-seq, two NGS-based methods, enable the study of 3D chromatin organization and interactions. These techniques enable the creation of chromatin interaction maps by capturing distant chromatin interactions [33, 34]. Researchers may learn more about the spatial organization of the genome and how it affects gene regulation by merging 3D chromatin conformation data with epigenetic mutations, gene expression data, and functional annotations.

1.4.3.5 Epigenomics and transcriptomics integrated profiles

To clarify the relationship between epigenetic alterations and gene expression, NGS data from epigenomics may be integrated with transcriptomics data. Differentially methylated regions (DMRs) associated with variations in gene expression can be found by integrating DNA methylation profiles with RNA-seq data. Incorporating information on chromatin accessibility and histone modifications with RNA-seq enables the research of epigenetic regulatory processes and the discovery of regulatory components related with specific gene expression patterns.

1.4.4 Metagenomics

The field of metagenomics focuses on the direct genetic investigation of the prokaryotic genome, which includes bacteria, fungi, and viruses present in a sample using either a targeted method or an adaptor ligation PCR technique for shotgun sequencing in culture-independent manner of the targeted strategy [35]. The hypervariable region encompassing 16S or 18S ribosomal RNA genes of bacteria and fungi is applied. Each bacterial species in the sample may be identified using a combination of hypervariable and conserved regions. The ITS1 and ITS2 regions spanning the 5.8S rRNA gene of the fungal genome are chosen for amplification in a manner similar to that for identifying specific species of fungi [36]. In order to research viral diversity, abundance, and functional potential of viruses in the environment, readings obtained from NGS (shotgun) are once again the culture-independent technique.

1.5 How to Access Next–Generation Sequencing Data?

All types of next-generation sequencing (NGS) analyses require access to public data. Familiarity with the associated tools and databases, as well as a working knowledge of the command line, are often needed to access this data. The Gene Expression Omnibus (GEO) website of the NCBI and the Sequence Read Archive repositories are helpful resources for searching openly accessible experimental sequencing data. For accessing reference data, several databases are there, i.e., the Ensembl database, iGenomes, and Wormbase. The Cancer Genome Atlas Project (TCGA), a National Cancer Institute (NCI)-funded consortium project, is also a source of NGS datasets including whole-genome sequence (WGS), whole-exome sequence (WXS), methylation, RNA expression, proteomic, and clinical datasets. We can access TCGA datasets from Genomic Data Commons (GDC) Data Portal.

1.6 Role of NGS in Research

1.6.1 Microbiome research

Microbes are everywhere around us; thus, their symbiotic, pathogenic, and commensal features are important to humans because they enable us to create an environment that works effectively. Through evolution, our ability to survive depends on the microbiome community. To be healthy, however, it is essential to closely monitor and fully understand host–microbiome and microbiome–intercommunity interactions. Pathogen surveillance, functional dysbiosis, and therapeutic potential are the methods. In order to assess the functional potential of the microbiome, genetic studies have connected the gut microbiome to disorders affecting mental health, autoimmune diseases (such as rheumatoid arthritis), and metabolic disorders (such as diabetes and obesity). This creates room for more therapy modalities and choices. Finding infections with known antibiotic resistance can be aided by designing focused panels to pick up mutations (helping in antibiotic resistance tracking) or by identifying the pathogenic genes followed by sequencing. Additionally, studies into the pharmacomicrobiomics of patients in need of medication are in progress. This would make it easier to determine how medications affect one's microbiome and how the microbiome handles drugs [3].

1.6.2 Human disease research

The study of the transcriptome, epi-transcriptome, and epigenome is currently a primary objective of NGS-based research in addition to genomic research. The study of human diseases caused by new pathogenic/emerging variants, such as SARS-CoV-2 variants, has also been done. Human genome-based research through WGS and WES has provided novel insights into biological processes and found application in wellness research, agriculture and food research, genome-wide association research studies uncovering the wide range of population genetic variants, their genetic linkage, and molecular basis to various diseases, including cancer. The primary areas of research where RNA-seq is used include the study of pathogen transcriptomic signatures, metastatic biomarkers, treatment resistance, immunological microenvironment, immunotherapy, and neoantigen studies in cancer. Through the use of single-cell RNA-sequencing techniques along with NGS, it is now feasible to investigate single-cell behavior with regard to its differentiation, de-differentiation, proliferation, and carcinogenesis in cancer. Epigenetic signatures are significant biomarkers that have significance not only for cancer, malignant transformation, and metastasis but also for their clinical applicability in other disease situations like diabetes, neurological diseases, infectious diseases, and immunological disorders. Subsequently, a large number of lncRNAs are discovered to be expressed differently in a variety of tumors [23]. As their aberrant abundance has been correlated to the stage of disease and prognosis, more studies are showing miRNAs as biomarkers for multiple cancers [3].

1.7 What is Pattern Recognition?

Pattern recognition is a field of study that involves the automated identification and classification of patterns or regularities within data [37]. It is a branch of artificial intelligence and machine learning that focuses on developing algorithms and techniques to recognize and interpret patterns in various forms of data, such as images, signals, text, or numerical data.

The goal of pattern recognition is to extract meaningful information and identify structures or relationships within the data that may not be readily apparent to human observers [38]. It involves both the design and implementation of algorithms that can learn from training data and make predictions or classifications on new, unseen data.

Pattern recognition techniques can be broadly categorized into two major groups: supervised and unsupervised learning [39]. In supervised learning, a model is trained using labeled data, where each pattern, sample, or instance is associated with a known class label or category. The model seeks to learn in recognizing patterns from the labeled data and can then classify new, unlabeled patterns. In unsupervised learning, on the other hand, the model works with unlabeled data and aims to discover inherent patterns or structures within the data without any predefined classes.

Pattern recognition has numerous applications across various domains. Some examples include:

Computer vision: Identifying objects, recognizing faces, and detecting anomalies or objects of interest in images or videos.

Speech recognition: Converting spoken language into written text.

Natural language processing: Analyzing and understanding text, including sentiment analysis, text classification, and information extraction.

Biometrics: Recognizing and verifying individuals based on their unique physiological or behavioral characteristics, such as fingerprint or voice recognition.

Financial data analysis: Identifying patterns or trends in financial markets or predicting stock price movements.

Medical diagnosis: Assisting in the interpretation of medical images, such as X-rays or MRI scans, or identifying disease patterns from patient data.

Overall, pattern recognition plays a crucial role in extracting valuable insights and making informed decisions from complex and large-scale data in various domains.

1.8 Basics of Pattern Recognition

The recognition of patterns in data is a fundamental notion within the field of artificial intelligence and machine learning that intends to recognize patterns and regularities in data. Through recognition of patterns, it enables machines to perceive and understand their surrounding environment, enabling them to make informed decisions. In this section, the key aspects of a pattern recognition system are discussed, including its definition, scope, applications, and components of a pattern recognition system.

1.8.1 Definition and scope

Pattern recognition is the process of categorizing input data based on their inherent characteristics into predefined categories or classes. In this process, data is analyzed to uncover meaningful patterns that can be utilized for a variety of purposes, including classification, clustering, object recognition, and anomaly detection. Pattern recognition extends across various domains, including computer vision, speech recognition, natural language processing, medical diagnosis, financial analysis, and more.

Thc scope of pattern recognition encompasses both supervised and unsupervised learning paradigms. In supervised learning, the system learns from labeled data, where the correct classes of patterns are provided. In unsupervised learning, the system explores the data's inherent structures without predefined labels, making it suitable for tasks like clustering.

1.8.2 Applications and importance

Pattern recognition finds applications in numerous real-world scenarios, significantly enhancing automation and decision-making processes. Some of its key applications include:

Image and object recognition: Pattern recognition is extensively used in computer vision to identify and classify objects within images and videos. It enables applications such as facial recognition, object tracking, and autonomous vehicle navigation.

Speech and audio processing: Speech recognition systems use pattern recognition techniques to convert spoken language into text, facilitating voice assistants, transcription services, and more.

Natural language processing: Pattern recognition helps in sentiment analysis, namely entity recognition, language translation, and other tasks involving understanding and processing human language.

Medical diagnosis: Pattern recognition aids in analyzing medical images (like X-rays and MRI scans) for disease detection, assisting clinicians in accurate diagnoses.

Financial analysis: It plays a role in predicting stock market trends, credit scoring, fraud detection, and other financial analytics.

Biometrics: Pattern recognition underlies biometric systems that identify individuals based on unique physiological or behavioral traits, such as fingerprints, iris scans, and gait analysis.

1.8.3 Components of a pattern recognition system

A pattern recognition system typically consists of several interconnected components working together to process and categorize data:

Data acquisition: This component involves capturing and digitizing raw data, such as images, audio, or text, to make it suitable for processing.

Feature extraction: Features are distinctive attributes or characteristics that represent patterns in the data. This step involves selecting relevant features that aid in discrimination between different classes.

Preprocessing: Data preprocessing includes tasks like normalization, noise reduction, and data augmentation to enhance the quality of input data.

Classification/clustering algorithm: This core component applies learning algorithms to classify data into predefined classes (classification) or discover inherent structures (clustering).

Training and learning: In supervised learning, the system is trained using labeled data to learn the relationships between features and classes. Unsupervised learning involves exploring data patterns without explicit labels.

Validation and testing: After training, the system is validated and tested using separate datasets to assess its performance, generalization, and potential overfitting.

Post-processing: Post-processing steps may involve refining classification results, handling outliers, or making decisions based on confidence scores.

Decision–making: The final output of the pattern recognition system is the decision or classification result based on the input data and the learned model.

Understanding the basics of pattern recognition and its underlying components is essential for designing and developing effective pattern recognition systems that can address a wide range of practical challenges and applications.

1.9 Feature Extraction and Representation

In pattern recognition, feature extraction and representation are pivotal steps that transform raw data into a more manageable and informative format, facilitating effective analysis and classification. This section delves into the concepts of feature extraction, the various types of features, and dimensionality reduction techniques that play a crucial role in enhancing the efficiency and accuracy of pattern recognition systems.

1.9.1 Feature selection vs. feature extraction

Feature selection and feature extraction are two fundamental strategies employed to enhance the quality of data representation [40]:

Feature selection: This approach involves identifying and retaining a subset of the most relevant features from the original data while discarding less informative ones. It aims to reduce the complexity of the data representation and improve the efficiency of subsequent processing. Feature selection methods include filters (statistical measures), wrappers (using the performance of a classifier), and embedded methods (features selected during model training).

Feature extraction: Feature extraction transforms the original data into a new set of features by combining or projecting the existing features. The goal is to create a more compact and meaningful representation while preserving as much relevant information as possible [40]. Techniques like principal component analysis (PCA), linear discriminant analysis (LDA), and independent component analysis (ICA) are commonly used for feature extraction.

1.9.2 Types of features

Different types of features capture different characteristics of data, and their suitable selection is determined by the pattern recognition task at hand:

Spatial features: The spatial arrangement of pixels in an image is referred to as spatial features. They record data regarding patterns, forms, and structures. Edge detection, corner detection, and region-based features, for example, are common spatial features used in image analysis.

Spectral features: Spectral features are extensively employed in spectral-related applications such as remote sensing and medical imaging. These qualities record information about the intensity of light at various wavelengths, allowing materials or tissues to be distinguished depending on their spectral properties.

Textural features: Textural features characterize the texture patterns in a picture, exhibiting properties such as smoothness, roughness, and coarseness. Textural properties are quantified using statistical measurements such as entropy, contrast, and homogeneity.

Temporal features: In time-series data analysis, temporal features capture patterns throughout time. They are critical for applications such as speech recognition, where the dynamics of the signal provide vital information.

1.9.3 Dimensionality reduction techniques

High-dimensional data can introduce computational challenges and boost the likelihood of overfitting. Dimensionality reduction approaches aim to deal with these issues by lowering the numbers of feature while retaining as much essential information as possible. By doing so, these strategies not only improve computational efficiency but also assist in preserving the intrinsic properties of the data:

Principal component analysis (PCA): PCA reorients data into a new coordinate system in which the principal components, acting as new axes, efficiently capture the most variance [41]. This method allows for the decrease of data dimensions while retaining the majority of the initial variance.

Linear discriminant analysis (LDA): LDA is typically employed for classification jobs. It seeks linear groupings of features that enhance the distinction between classes while lowering the variability within each of the classes [42].

t-Distributed stochastic neighbor embedding (t-SNE): t-SNE is particularly useful for visualizing high-dimensional data in lower-dimensional spaces, often used for data exploration and visualization.

Autoencoders: Autoencoders are neural network topologies employed for unsupervised learning. They are made up of an encoder and a decoder, and their goal is to reconstruct the input data by compressing it into a lower-dimensional latent space.

Dimensionality reduction techniques not only help mitigate the "curse of dimensionality" but also aid in visualizing data, speeding up processing, and improving the generalization of pattern recognition models.

In summary, feature extraction and representation are integral to effective pattern recognition. They allow for the transformation of raw data into more informative and manageable forms, enhancing the performance and interpretability of pattern recognition systems across diverse domains.

1.10 Similarity and Distance Metrics

In the realm of pattern recognition and data analysis, measuring similarity or dissimilarity between data points is crucial for various tasks such as clustering, classification, and recommendation systems. Different distance and similarity metrics help quantify the relationships between data points in multidimensional spaces. This section explores some widely used metrics for measuring similarity and distance.

1.10.1 Euclidean distance

Euclidean distance is one of the most common distance metrics used to measure the straight-line distance between two points in a multidimensional space. It is calculated using the Pythagorean theorem and is suitable for continuous data. The formula for Euclidean distance between two points A and B with *n* dimensions is:

$$\sqrt{\sum_{i=1}^{n}(A_i - B_i)}.$$

1.11 What are Features, Classes, and Dataset?

In the context of pattern recognition and machine learning, let us define the terms "features," "classes," and "dataset":

Features: Features refer to the measurable properties or characteristics of the data that are used to describe or represent the patterns. They are derived from the input data and capture relevant information that helps in distinguishing different patterns or instances. Features can be numerical, categorical, or even derived from more complex data types like images or text. For example, if you are working with images of animals, the features could include the color distribution, texture, shape, or any other measurable property that differentiates one animal from another. These features are typically represented as a set of numerical values or a feature vector.

Classes: Classes, also known as categories or labels, represent the different categories or groups to which patterns or instances belong. In supervised learning, classes are predefined and known during the training phase [43]. The goal of pattern recognition is to assign each instance to one of these classes based on its features. For instance, if you are working on a classification task for email spam detection, the two classes could be "spam" and "non-spam" or "ham." The pattern recognition system aims to classify each email as either spam or non-spam based on its features.

Dataset: A dataset is a collection of instances or examples that are used for training, testing, or evaluating a pattern recognition system. It consists of input data along with their corresponding labels or class assignments.

A dataset may include both the input data and their associated features. In supervised learning, the dataset is typically divided into two subsets: the training set and the test set. The training set is used to train the pattern recognition system by providing labeled examples, while the test set is used to evaluate the system's performance on unseen data.

The dataset serves as the foundation for building and evaluating pattern recognition models. It provides the necessary information for learning patterns from data, extracting features, and training the system to make accurate predictions or classifications on new, unseen instances.

It is worth noting that these terms are interconnected and essential in the context of pattern recognition and machine learning, as they form the basis for understanding, representing, and analyzing data to recognize patterns and make predictions or classifications [39].

1.12 Why Is Pattern Recognition Important?

Pattern recognition is important for several reasons:

Data analysis and interpretation: In today's world, we generate vast amounts of data in various forms. Pattern recognition enables us to analyze and interpret this data effectively. By recognizing patterns, we can identify trends, correlations, and anomalies within the data, leading to valuable insights and informed decision-making.

Automation and efficiency: Pattern recognition plays a crucial role in automating tasks that would otherwise be time-consuming or labor-intensive for humans. It enables machines to automatically process and classify data, reducing manual effort and increasing efficiency in tasks such as image or speech recognition, text analysis, and data mining.

Predictive modeling: By recognizing patterns in historical data, pattern recognition techniques can be used to build predictive models. These models can then be utilized to forecast future trends, make predictions, or estimate outcomes in various domains, including finance, healthcare, weather forecasting, and marketing.

Anomaly detection: Pattern recognition helps identify anomalies or outliers within datasets. By understanding the normal patterns and structures, any deviations or abnormalities can be detected and flagged for further investigation. This is valuable in detecting fraud, network intrusions, equipment failures, and other irregularities.

Decision support: Pattern recognition systems provide decision support by analyzing and summarizing complex data. By recognizing patterns, the systems can provide insights, recommendations, or predictions to assist decision-makers in making more informed choices.

Biometrics and security: Pattern recognition techniques are extensively used in biometric systems for identification and verification. They enable recognition of unique patterns in individual traits such as fingerprints, iris scans, facial features, or voice patterns, providing secure authentication and access control in applications like passport verification, secure banking, or building access.

Medical diagnosis and healthcare: Pattern recognition is utilized in medical imaging and diagnosis, aiding in the interpretation of medical images and identifying patterns associated with diseases or conditions. It helps in early detection, treatment planning, and personalized medicine.

Overall, pattern recognition is important because it enables us to extract knowledge and insights from complex data, automate tasks, make predictions, detect anomalies, and support decision-making across various domains. It empowers machines to process and understand information in ways that mimic human cognitive abilities, contributing to advancements in technology, science, and everyday life.

1.13 What Is a Pattern Recognition System?

A pattern recognition system is a computational system or framework that is designed to recognize and interpret patterns within data. It consists of algorithms, models, and processes that enable the system to analyze input data, extract meaningful features, and make decisions or classifications based on the identified patterns.

The components of a pattern recognition system typically include:

Input data: This can be in various forms such as images, signals, text, or numerical data. The input data serves as the raw information on which the pattern recognition system operates.

Feature extraction: In order to recognize patterns, the system needs to extract relevant features or characteristics from the input data. Feature extraction techniques are applied to transform the raw data into a more concise and representative form that captures important information for pattern recognition.

Classification or decision-making: Once the features are extracted, the pattern recognition system employs classification algorithms or decision-making mechanisms to assign the input data to specific classes or categories.

This step involves using the learned patterns from training data to make predictions or decisions on new, unseen data.

Training and learning: Many pattern recognition systems employ machine learning techniques to learn from labeled training data. During the training phase, the system learns to recognize patterns by adjusting its internal parameters or model based on the provided training examples. This learning process enables the system to generalize and make accurate predictions on new, unseen data.

Evaluation and refinement: After the system has been trained, it needs to be evaluated and refined to ensure its performance and accuracy. This involves testing the system on a separate set of data called the test set and measuring its performance metrics, such as accuracy, precision, recall, or F1 score. The system can be further refined by adjusting its parameters, improving feature extraction methods, or incorporating more diverse training data.

Pattern recognition systems find applications in a wide range of fields, including computer vision, speech recognition, natural language processing, biometrics, and many others. They enable machines to automatically analyze and interpret data, recognize complex patterns, and make intelligent decisions based on the identified patterns.

1.14 Biomedical Imaging

Biomedical images are scaled measurements of the human body (such as microscopic, macroscopic, etc.). They can measure a physical characteristic of the human body, such as radio-density, or the opacity to X-rays, and come in a wide range of imaging modalities (such as a CT scanner, an ultrasound machine, etc.). These images are analyzed by specialists in the field (such as radiologists) for clinical activities (such as a diagnosis) and have a significant impact on doctors' decisions. Imaging is the visual depiction of structures and functions of objects including biological molecules, biological ultrastructures, tissues, and the spatial organization of the item. Additionally, it is a critical stage in contemporary medicine's diagnostic and treatment processes. For instance, CT scans have been employed as a primary diagnostic standard during the current COVID-19 pandemic in addition to nucleic acid testing. In contrast to computers, humans are far better at understanding and interpreting information obtained from image data than they are at doing the same with information derived from numerical or textual data. On the other side, enhanced automation of AI technologies may result in more objective and highly reproducible analytical outcomes. Therefore, it is advantageous to employ AI techniques to support manual image analysis. Both biological and biomedical sciences are relying more and more on imaging. Biological imaging has produced rich information about biological systems and molecules at different resolutions, all the way from tissue-level to cellular-level to organelle-level to macro molecular-level to small-molecular-level to atomic-level, using technologies like optical microscopy, fluorescence microscopy, electron tomography, nuclear magnetic resonance, single particle cryo-EM, and X-ray crystallography. Different imaging modalities, including ultrasound, computed tomography (CT), magnetic resonance imaging (MRI), positron emission tomography (PET), and optical coherence tomography (OCT), have a wide range of diagnostic and therapeutic uses in medicine. Such technologies can give clinicians and doctors quick, non-invasive, painless, and direct information that is essential for diagnosis, prognosis, and therapy [44].

1.15 Medical Imaging Types and Modalities

There are several different types of medical imaging, and as technology progresses, more methods are being developed. Each type uses a different technique to generate images of what is happening inside the body [45].

1.15.1 Magnetic resonance imaging (MRI)

In order to view the organs and other structures in the human body, this imaging technique uses magnetic fields and radio waves. An MRI scanner is required for the procedure, which is a giant tube that holds a significant circular magnet. The protons of the hydrogen atoms in the body are aligned by the magnetic field produced by this magnet. Subsequently, the protons are subjected to radio waves, which causes them to rotate. The protons relax and realign themselves once the radio waves turn off, producing radio waves in the recovery process that can be detected by the system to develop an image.

1.15.2 Ultrasound

To generate images of joints, muscles, organs, and soft tissues, this imaging technique uses high-frequency sound waves that are reflected off tissue. It resembles flashing a light into the body, but the light passes through the skin layers and can only be seen with the aid of electronic sensors. Ultrasound is one of the safest and most cost-effective medical imaging methods available. It also offers a wide variety of uses and has no negative side effects.

1.15.3 Radiography

The interior of the body is captured using electromagnetic radiation during this imaging method. X-rays are the most widely used and common form of radiography. An X-ray machine uses high-energy radiation to image the body during this operation. Organs and skin are examples of soft tissues that do not absorb these waves, but bones and other hard tissues do. The device prints the X-ray outcomes, highlighting the bodily sections that absorbed the waves in white and leaving the unabsorbed material in black.

1.15.4 Computer tomography (CT)

These are an X-ray type that creates 3D images to aid in diagnosis. It is also known as the name computed axial tomography (CAT), and it creates cross-sectional images of the human body using X-rays. The patient may lie on a motorized table inside the enormous circular hole of the scanner. A narrow, "fan-shaped" X-ray beam is then created by rotating the detector and X-ray around the patient, passing through a portion of the patient's body to develop an image. Compared to traditional X-rays, CT scans provide clearer, more accurate images of the body's bones, blood arteries, internal organs, and soft tissue. The use of CT scans typically prevents the need for exploratory surgery.

1.15.5 Positron emission tomography (PET)

Positron emission tomography (PET) [46] is a method for assessing physiological function by examining neurotransmitters, blood flow, metabolism, and radiolabeled medicines. PET provides quantitative studies that make it possible to track relative changes over time as a disease process grows or in response to a specific stimulus.

1.15.6 Optical coherence tomography (OCT)

A non-contact imaging method called optical coherence tomography (OCT) creates high-resolution cross-sectional images of tissue. It is therefore especially useful for organs, such as the human eye, where standard microscopic tissue identification by biopsy is not possible. OCT delivers *in vivo* images without affecting the tissue being imaged since it is entirely noninvasive. An image may be visualized in real time and at video rate for rapid scanning speeds and signal processing. Compared to other medical imaging techniques like ultrasound or magnetic resonance imaging (MRI), OCT has a far greater resolution. It combines lateral resolution similar to confocal scanning laser ophthalmoscopy with an axial resolution that can match that of confocal microscopy [47].

1.16 AI-based Biomarker Diagnosis from Medical Images

We may define biological target volume with the use of artificial intelligence (AI) tools by detecting biomarkers in functional images. Methods for classifying different tumors can be categorized as manual, semi-automated, or completely automatic depending on the degree of interaction from users. Radiologists employ multi-modality data from the medical images as well as knowledge of anatomy and physiology for the manual segmentation procedures. For this type of segmentation, the radiologist must carefully manually sketch the regions and areas of the tumor by navigating through CT/MRI scan slices. As a result, manual segmentation is a laborious, radiologist-dependent procedure with unpredictable outcomes. However, manual segmentations can be helpful for evaluating the outcomes of completely automatic and semi-automatic approaches [48].

Three procedures are often required for semi-automatic cancer imaging segmentation methods: initialization, intervention response or feedback reaction, and outcome evaluation. In order to enable the automated algorithm approach, initialization is often accomplished through the process of identifying the region of interest (ROI), which contains the approximate region of the tumor. Then, by making alterations depending on the feedback acquired in response, machine learning algorithms may be employed to get an optimal output during the process. Users, including doctors and physicists, can evaluate the outcome and change or repeat the procedure if the outcomes are unacceptable. On medical images, the user must draw a circle around the tumor's maximal diameter using the semi-automatic segmentation technique known as tumor cut. The segmentation problem was transformed into a classification challenge in an enhanced tumor-cut semi-automatic technique, which is proposed by Sompong and Wongthanavasu [49]. For tumor segmentation, machine learning classification algorithms need a lot of data in the form of imaging scans (with known ground truth) from many different subjects that have been corrected for intensity bias and noise. Rather, we may utilize a subset of voxels from each type of tissue. The approach extracts the intensity values and spatial coordinates from these chosen voxels as features, and then trains a support vector machine (SVM) to classify all the voxels of the same input image to the appropriate tissue type [49]. Cancer diagnosis with semi-automatic segmentation takes less time than manual approaches and can produce effective outcomes; however, semi-automatic tumor segmentation techniques are still subject to intra- and inter-user variability. As a result, completely automated techniques are the main focus of current tumor segmentation research.

Fully automated tumor segmentation approaches do not require user intervention. For some tumor types, such as gliomas, automated segmentation can be quite difficult. Tumor forms, sizes, and locations are highly variable among patients in 3D tumor-bearing imaging data. Additionally, the tumor borders are typically inconsistent and ambiguous with discontinuities, which presents a significant problem, particularly for edge-based techniques. For instance, MRI images of brain tumors from clinical scans or artificial databases are inherently complicated. Each image in the dataset has intensity biases and other variances as a result of the vast differences between the MRI machines and data-obtaining techniques. Complexity is even increased by the requirement for many modalities to segregate tumor sub-regions efficiently. While innovative completely automatic tumor segmentation approaches based on deep learning algorithms are also developing with excellent results, the manual and semi-automatic classification methods continue to show significant performances. The field of cancer diagnostics has seen extensive use of AI.

In this book, Chapter 2 presents a framework for integrating DNA methylation and associated TCGA gene expression data to uterine cervical cancer utilizing linear regression, differential expression, and deep learning. The outcomes of our framework proved that it is more beneficial for improving biological interpretation of the DNA methylation data than just performing a single differential methylation study or identifying differentially methylated regions. In Chapter 3, we discuss the development of multi-objective optimized variable cutoff-based association rule mining ($MOOVARM$) for multi-omics profile based on minimum distance from positive ideal solution (PIS), and minimum distance from negative ideal solution (NIS). Any rule might be regarded as a final resultant rule if its relative closeness score exceeded and was equal to the predetermined cutoff value. We also created a ranking of the rules based on the relative score that can be used to provide improved disease classification performance compared to the most state-of-the-art algorithms in disease diagnosis and therapeutic response. Since dimensionality reduction and clustering have grown interest in recent years for the interpretation of single-cell RNA sequencing data, in Chapter 4, we present an integrated clustering model with dimensionality reduction for the identification of cluster-specific biomarkers in single-cell sequencing data. Our integrated framework effectively identifies overlapping biomarkers from single-cell RNA sequencing data as cluster-specific frequent biomarkers. In Chapter 5. we develop an expression-based predictor to identify the essential human genes. A novel feature extraction and analysis method has been taken into consideration for the classification of essential genes in the species and the detection of discriminant features. Chapter 6 represents an integrated model of deep learning and next-generation medical image data for rare disease stage detection. The proposed CNN model is trained and tested using a dataset of brain MRI images, and the same model is also applied to the SHAP and LIME methods to provide an explanation. For better feature extraction in our experiment, we used two dataset copies as input — one in the convolution layer and another in the fully connected layer. The overall performance of the prediction model was negatively affected by any attempt to eliminate any features, and hence no augmentation was made. The proposed framework is a locally interpreted model with a model-agnostic explanation that is clearly stated to more qualitatively depict the outcomes for typical individuals. Chapter 7 concludes all of our developed frameworks and shows the future research direction of this domain.

References

[1] https://www.thermofisher.com/in/en/home/life-science/sequencing/sequencing-learning-center/next-generation-sequencing-information/ngs-basics/what-is-next-generation-sequencing.html

[2] Qin D. Next-generation sequencing and its clinical application. Cancer Biol Med. 2019 Feb;16(1):4-10. doi: 10.20892/j.issn.2095-3941.2018.0055. PMID: 31119042; PMCID: PMC6528456.

[3] Satam H, Joshi K, Mangrolia U, Waghoo S, Zaidi G, Rawool S, Thakare RP, Banday S, Mishra AK, Das G, et al. Next-Generation Sequencing Technology: Current Trends and Advancements. Biology. 2023; 12(7):997. https://doi.org/10.3390/biology12070997

[4] Costain, G.; Cohn, R.D.; Scherer, S.W.; Marshall, C.R. Genome sequencing as a diagnostic test. Can. Med. Assoc. J. 2021, 193, E1626–E1629.

[5] Rabbani, B.; Tekin, M.; Mahdieh, N. The promise of whole-exome sequencing in medical genetics. J. Hum. Genet. 2014, 59, 5–15.

[6] Iglesias, A.; Anyane-Yeboa, K.; Wynn, J.; Wilson, A.; Cho, M.T.; Guzman, E.; Sisson, R.; Egan, C.; Chung, W.K. The usefulness of whole-exome sequencing in routine clinical practice. Anesth. Analg. 2014, 16, 922–931.

[7] Van Dijk, E.L.; Auger, H.; Jaszczyszyn, Y.; Thermes, C. Ten years of next-generation sequencing technology. Trends Genet. 2014, 30, 418–426.

[8] Warr, A.; Robert, C.; Hume, D.; Archibald, A.; Deeb, N.; Watson, M. Exome Sequencing: Current and Future Perspectives. G3 Genes Genom. Genet. 2015, 5, 1543–1550.

[9] Williams, M.J.; Sottoriva, A.; Graham, T.A. Measuring Clonal Evolution in Cancer with Genomics. Annu. Rev. Genom. Hum. Genet. 2019, 20, 309–329.

[10] Li, J.; Liu, C. Coding or Noncoding, the Converging Concepts of RNAs. Front. Genet. 2019, 10, 496.

[11] Lucchinetti, E.; Zaugg, M. RNA Sequencing. Anesthesiology 2020, 133, 976–978.

[12] Choi, S.-W.; Kim, H.-W.; Nam, J.-W. The small peptide world in long noncoding RNAs. Brief. Bioinform. 2019, 20, 1853–1864.

[13] Lasda, E.; Parker, R. Circular RNAs: Diversity of form and function. RNA 2014, 20, 1829–1842.

[14] Chen, J.-W.; Shrestha, L.; Green, G.; Leier, A.; Marquez-Lago, T.T. The hitchhikers' guide to RNA sequencing and functional analysis. Brief. Bioinform. 2023, 24, bbac529.

[15] Stark, R.; Grzelak, M.; Hadfield, J. RNA sequencing: The teenage years. Nat. Rev. Genet. 2019, 20, 631–656.

[16] Ura, H.; Togi, S.; Niida, Y. A comparison of mRNA sequencing (RNA-Seq) library preparation methods for transcriptome analysis. BMC Genom. 2022, 23, 303.

[17] Kolanowska, M.; Kubiak, A.; Jażdżewski, K.; Wójcicka, A. MicroRNA Analysis Using Next-Generation Sequencing. Methods Mol. Biol. 2018, 1823, 87–101.

[18] Grillone, K.; Riillo, C.; Scionti, F.; Rocca, R.; Tradigo, G.; Guzzi, P.H.; Alcaro, S.; Di Martino, M.T.; Tagliaferri, P.; Tassone, P. Non-coding RNAs in cancer: Platforms and strategies for investigating the genomic "dark matter". J. Exp. Clin. Cancer Res. 2020, 39, 117.

[19] Atkinson, S.R.; Marguerat, S.; Bähler, J. Exploring long non-coding RNAs through sequencing. Semin. Cell Dev. Biol. 2012, 23, 200–205.

[20] Raghavan, V.; Kraft, L.; Mesny, F.; Rigerte, L. A simple guide to de novo transcriptome assembly and annotation. Brief. Bioinform. 2022, 23, bbab563.

[21] Kulkarni, A.; Anderson, A.G.; Merullo, D.P.; Konopka, G. Beyond bulk: A review of single cell transcriptomics methodologies and applications. Curr. Opin. Biotechnol. 2019, 58, 129–136.

[22] Adil, A.; Kumar, V.; Jan, A.T.; Asger, M. Single-Cell Transcriptomics: Current Methods and Challenges in Data Acquisition and Analysis. Front. Neurosci. 2021, 15, 591122.

[23] Wang, J.; Tian, T.; Li, X.; Zhang, Y. Noncoding RNAs Emerging as Drugs or Drug Targets: Their Chemical Modification, Bio-Conjugation and Intracellular Regulation. Molecules 2022, 27, 6717.

[24] López-Camarillo, C.; Gallardo-Rincón, D.; Álvarez-Sánchez, M.E.; Marchat, L.A. Pharmaco-epigenomics: On the Road of Translation Medicine. In Translational Research and Onco-Omics Applications in the Era of Cancer Personal Genomics; Springer: Berlin/Heidelberg, Germany, 2019; Volume 1168, pp. 31–42.

[25] Handy, D.E.; Castro, R.; Loscalzo, J. Epigenetic Modifications. Circulation 2011, 123, 2145–2156.

[26] Barros-Silva, D.; Marques, C.J.; Henrique, R.; Jerónimo, C. Profiling DNA Methylation Based on Next-Generation Sequencing Approaches: New Insights and Clinical Applications. Genes 2018, 9, 429.

[27] Lu, R.J.-H.; Liu, Y.-T.; Huang, C.W.; Yen, M.-R.; Lin, C.-Y.; Chen, P.-Y. ATACgraph: Profiling Genome-Wide Chromatin Accessibility from ATAC-seq. Front. Genet. 2021, 11, 618478.

[28] Mansisidor, A.R.; Risca, V.I. Chromatin accessibility: Methods, mechanisms, and biological insights. Nucleus 2022, 13, 238–278.

[29] Liu, E.T.; Pott, S.; Huss, M. Q&A: ChIP-seq technologies and the study of gene regulation. BMC Biol. 2010, 8, 56.

[30] Furey, T.S. ChIP-Seq and beyond: New and improved methodologies to detect and characterize protein-DNA interactions. Nat. Rev. Genet. 2012, 13, 840–852.

[31] O'geen, H.; Echipare, L.; Farnham, P.J. Using ChIP-Seq Technology to Generate High-Resolution Profiles of Histone Modifications. Methods Mol. Biol. 2011, 791, 265–286.

[32] Nakato, R.; Sakata, T. Methods for ChIP-seq analysis: A practical workflow and advanced applications. Methods 2021, 187, 44–53.

[33] Feng, F.; Yao, Y.; Wang, X.Q.D.; Zhang, X.; Liu, J. Connecting high-resolution 3D chromatin organization with epigenomics. Nat. Commun. 2022, 13, 2054.
[34] Tang, B.; Cheng, X.; Xi, Y.; Chen, Z.; Zhou, Y.; Jin, V.X. Advances in Genomic Profiling and Analysis of 3D Chromatin Structure and Interaction. Genes 2017, 8, 223.
[35] Thomas, T.; Gilbert, J.; Meyer, F. Metagenomics-A guide from sampling to data analysis. Microb. Inform. Exp. 2012, 2, 3.
[36] Bellemain, E.; Carlsen, T.; Brochmann, C.; Coissac, E.; Taberlet, P.; Kauserud, H. ITS as an environmental DNA barcode for fungi: An in silico approach reveals potential PCR biases. BMC Microbiol. 2010, 10, 189.
[37] Theodoridis, Sergios and Koutroumbas, Konstantinos, Pattern recognition, Book, 2006, Elsevier.
[38] Duda, RO and Hart, PE and Stork, DG and Ionescu, Alexandru, Book: Pattern classification, Chapter: nonparametric techniques, 2000, Wiley-Interscience Publication.
[39] Bishop, Christopher M and Nasrabadi, Nasser M, Pattern recognition and machine learning, volume 4, number 4, 2006, Springer.
[40] Guyon, Isabelle and Elisseeff, André, An introduction to variable and feature selection, Journal of machine learning research, volume 3, pages 1157–1182, Mar 2003.
[41] Jolliffe, I, Principal component analysis: Wiley online library, Google Scholar, 2005
[42] Fisher, Ronald Aylmer and others, 138: The Use of Multiple Measurements in Taxonomic Problems, 1936
[43] Hastie, Trevor and Tibshirani, Robert and Friedman, Jerome H and Friedman, Jerome H, The elements of statistical learning: data mining, inference, and prediction, book, volume 2, 2009, Springer.
[44] Gao X, Li L and Xu M (2021) Editorial: AI in Biological and Biomedical Imaging. Front. Mol. Biosci. 8:804476. doi: 10.3389/fmolb.2021.804476.
[45] https://www.postdicom.com/en/blog/medical-imaging-types-and-modalities
[46] Berger A. How does it work? Positron emission tomography. BMJ. 2003 Jun 28;326(7404):1449. doi: 10.1136/bmj.326.7404.1449. PMID: 12829560; PMCID: PMC1126321.
[47] Aumann S, Donner S, Fischer J, et al. Optical Coherence Tomography (OCT): Principle and Technical Realization. 2019 Aug 14. In: Bille JF,

editor. High Resolution Imaging in Microscopy and Ophthalmology: New Frontiers in Biomedical Optics [Internet]. Cham (CH): Springer; 2019. Chapter 3. Available from: https://www.ncbi.nlm.nih.gov/books/NBK554044/doi:10.1007/978-3-030-16638-0_3

[48] Pang Y, Wang H, Li H., Medical Imaging Biomarker Discovery and Integration Towards AI-Based Personalized Radiotherapy, Front Oncol. 2022 Jan 17;11:764665. doi: 10.3389/fonc.2021.764665. PMID: 35111666; PMCID: PMC8801459.

[49] Sompong C, Wongthanavasu S. An Efficient Brain Tumor Segmentation Based on Cellular Automata and Improved Tumor-Cut Algorithm. In: Expert Systems With Applications, vol. 72. Elservier (2017). p. 231–44.

2

Integrating DNA Methylation, Linear Regression, and Machine Learning on RNA-seq Data

2.1 Introduction

A promising biomarker for cancer diagnosis and classification has been discovered: DNA methylation. DNA methylation is a heritable epigenetic mark that results from the covalent transfer of a methyl group to the C-5 position of the DNA cytosine ring by enzymes called DNA methyltransferases (DNMTs). DNA methylation is essential for healthy growth. It plays a crucial role in a number of vital processes, such as genomic imprinting, X chromosomal inactivation, the inhibition of repetitive element transcription and transposition, and several diseases, such as cancer [1]. Two types of analysis are provided to functionally assess the DNA methylation data: (i) single differentially methylated genes (CpG sites) discovery [2], [3] and (ii) differentially methylated region (DMR) finding [4], [5], [6]. These two categories of analysis are primarily designed to carry out a single function. Therefore, it is crucial to take into account a variety of factors when interpreting DNA methylation data so that it can serve multiple purposes, including the prediction of gene expression using DNA methylation, differential expression analysis, the classification of cancer, and hub gene discovery, among others.

Practical experience shows that DNA methylation often decreases the level of gene expression. However, depending on certain factors, this opinion changes. The methods used to combine data on DNA methylation and gene expression vary. These methods now in use have numerous drawbacks. First of all, it can be challenging to pinpoint the direction of the analyzed gene expression, which is inferred from DNA methylation. In most cases, hypermethylation in the promoter region suppresses gene expression, whereas hypermethylation in the gene body correlates with activation. As a result, it is challenging to detect changes in gene expression from simple DNA methylation findings [35]. Second, the variability in size of canonical promoters and the presence of distal augments make it difficult to detect gene promoter methylation accurately because these factors introduce biases into the correlation of methylated regions with gene models [35]. Thirdly, the likelihood of choosing a prolonged gene is increased by neighboring CpGs that have different levels of methylation or by the gene's overlap (or nesting) with others. [35]. Fourth, for some web-based approaches like GREAT [8] and Galaxy [9], some specialized tools are needed for reformatting the methylation data into the genomic region formats (for example, BED). It makes usage more difficult [35].

The cervix, a hollow tube that connects the lower portion of the uterus to a woman's vagina, is where cervical cancer first appears. The cells on the cervix's outer surface are where the majority of cervical malignancies develop. The symptoms are typically too similar to other frequent disorders including menstruation and urinary tract infections for women to recognize this disease in its early stages. Irregular bleeding during menstruation or after having intercourse, pelvic pain, and pain when urinating are all typical symptoms of cervical cancer [10]. In this study, we used a DNA methylation dataset for uterine cervical cancer from NCBI (Accession ID: GSE30760) [11], which includes samples from both normal and cancerous uteruses.

For accurate interpretation of DNA methylation in a complex disease like cancer, there is yet no method that combines regression, differential expression, and deep learning algorithms. In order to address these issues, we provided an integrated framework in this chapter [36], which integrates DNA methylation data with corresponding TCGA gene expression data for uterine cervical data in order to correctly interpret biologically the DNA methylation data using regression, differential expression, and deep learning methods (NCBI Accession ID: GSE30760) [11], [12], [13]. We pre-filtered the redundant CpG sites, removed outliers, and replaced missing values. After determining the relationship between DNA methylation and TCGA gene expression, we then estimated the associated gene expression value using a linear regression technique from the pre-filtered DNA methylation data. As a result, we were able to acquire the predicted gene expression matrix for the preprocessed DNA methylation data. We employed the ByMethyl R package for the entire analysis [35]. Then, we discovered DEGs by integrating downstream analysis, the empirical Bayes test with

Limma Voom [18], [19]. After interpreting those DEGs for the purpose of determining the classification ability of the uterine cancer and normal groups using the recently published deep learning method *nnet* (feed-forward neural network based model) [21], we estimated all classification metrics using 10-fold cross-validation, including average accuracy, average sensitivity, average specificity, average precision, and average overall error rate. With the default parameter settings for size (number of units in hidden layer), rang (initial random weights), decay (parameter for weight decay), maxit (maximum number of iterations or epochs), MaxNWts (maximum number of weights), and other default parameters, we trained our predicted DEG expression data using *nnet*. Significantly, using DEG expression data, we were able to classify uterine cervical cancer samples and normal samples with an average accuracy of $90.69\%(\pm 1.97\%)$. Comparative research shows that our suggested strategy has a higher classification accuracy than other state-of-the-art methods. Additionally, we used *Cytoscape* online tool (version 3.8.0) [14] to conduct in-degree and out-degree hub gene network analyses. We revealed the top 5 in-degree genes (*PAIP2, GRWD1, VPS4B, CRADD, and LLPH*) and the top 5 out-degree genes (*MRPL35, FAM177A1, STAT4, ASPSCR1, and FABP7*). Then, using *WebGestalt (WEB-based Gene SeT AnaLysis Toolkit)* [20], we carried out gene set enrichment analysis (GESA) to identify enriched KEGG pathways and gene ontology (GO) terms such as biological process (BP), cellular component (CC), and molecular function (MF) on the set of all DEGs with FDR< 0.001. Finally, compared to employing a single differential methylation analysis or methodologies for detecting differentially methylated regions for any type of cancer, our integrated framework using linear regression, differential expression, and deep learning can interpret the DNA methylation data better.

2.2 Materials and Methods

As well as in Figure 2.1, the following steps of our suggested framework are shown [36].

2.2.1 Data collection

A cervical cancer methylation dataset was used in this study (NCBI Accession ID: GSE30760) [11], [12], [13]. This dataset contained 152 matched normal samples and 63 samples of uterine cervical tumors. Of note, 27,578 genes were included in the initial analysis.

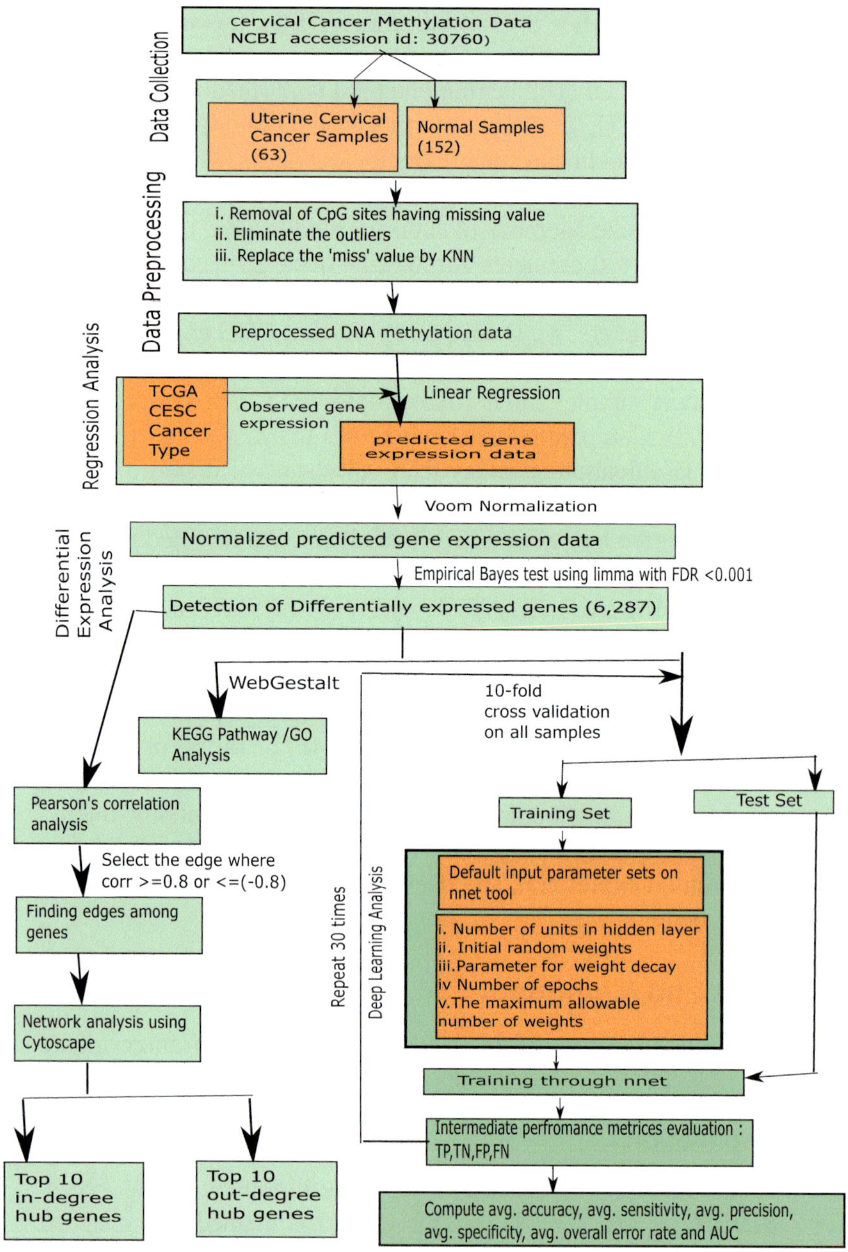

Figure 2.1 Flowchart of the proposed framework.

2.2.2 Preprocessing of methylation data

By using regression, differential expression, and deep learning, we performed a thorough study in this research to combine DNA methylation and related TCGA gene expression data. We have used various phases in this process, which are listed below.

2.2.2.1 Data preprocessing

The CpG sites with missing values in more than half of the samples were first deleted, and the remaining missing values were then imputed by integrating the new traditional quality control R package *ENmix* [15], which is extensively used in the analysis of Illumina Human Methylation data. The functions of *ENmix* package are used to reduce needless experimental noise and boost the accuracy and repeatability of methylation measures. The *ENmix* functions are quite flexible and precise. With the help of this quality control *ENmix*, we were able to restore missing values in our methylation data using the well-known k-nearest neighbors (KNN) algorithm and eliminate outliers in our proposed method.

2.2.2.2 Computing predicted expression scores of gene through linear regression analysis

In this phase, using preprocessed DNA methylation profiles and the related TCGA CESC cancer type coupled with a corresponding pre-trained weight factor, we derived the predicted gene expression scores. To do this, we measured the relationship between DNA methylation and gene expression for uterine cervical cancer on preprocessed DNA methylation and associated TCGA CESC cancer type using the linear regression algorithm [35]. From a statistical perspective, linear regression is a linear method for modeling the relationship between a scalar variable (or dependent variable) and one or more explanatory variables (or independent variables). DNA methylation (Mt) is the independent variable in a regression analysis, while gene expression (E_x) is its dependent variable. For i-indexed gene denoted by $gene_i$, $E_x = \{e_{x_1}, e_{x_2}, ..., e_{x_n}\}$ is the gene expression across n samples, and Mt is the corresponding methylation matrix ($27,578 \times 215$ matrix here). For the purpose of developing the model, we selected the CpGs ($cpg_{n,j}$) whose beta values were correlated, that is, whose Pearson's correlation coefficient was more than $|0.8|$ with the gene expression label ($gene_i$), where $cpg_{n,j}$ is the beta value of jth CpG in sample n. The following equation is an explanation of the linear regression model:

$$E_x = \alpha + beta * Mt, \tag{2.1}$$

where α stands for the linear regression intercepting factor, and $beta$ for the coefficient vector.

In our scenario, utilizing DNA methylation data and this linear regression model, we produced the predicted gene expression matrix for the given genes (CpG sites). Next, we used 10-fold cross-validation to check the accuracy of our model. This means that we must assess the accuracy of the gene expression that the linear regression model inferred. Basically, we used 9/10 samples as the training dataset to train the model for each validation. After that, we integrated the DNA methylation data and the trained model to compute a gene expression profile for the remaining 1/10 samples. After completing 10-fold cross-validation, the next step is to combine the gene expression profiles of all the test sample profiles into one. We matched the RNA-seq data with the gene expression to perform downstream validation.

2.2.2.3 Voom normalization and identifying differentially expressed genes using Limma

In this step, *Voom* normalization [17] is utilized, and then *Limma* [18], [14]. After employing *Voom* normalization tool, we extracted DEGs from the predicted gene expression data by *Limma* for downstream analysis [19]. For all types of data distributions and sample sizes, *Limma* performs quite well, according to benchmark methodologies.

The moderated t-statistic of *Limma* is defined as follows [19]:

$$\bar{t_k} = \left(\frac{1}{\sqrt{\left(\frac{1}{m_1}\right) + \left(\frac{1}{m_2}\right)}} \right) * \left(\frac{\hat{\beta_k}}{\bar{p_k}} \right), \tag{2.2}$$

where m_1 stands for the sample size of diseased group and m_2 indicates the sample size of the control group, and total sample size $m = m_1 + m_2$. $\hat{\beta_k}, p_k$ denotes the corresponding contrast estimator and posterior sample variance for the genes, respectively.

We used a t-table or cumulative distribution function to determine the false discovery rate (FDR) adjusted p-value using empirical Bayes t-statistics (cdf). The differentially expressed genes (DEGs) in this study are those with an FDR-adjusted p-value less than 0.001. This p-value is the likelihood that the specified null hypothesis will hold true when a t-value is either equal to or larger than the one that was actually observed.

Here, we implemented empirical Bayes test using *Limma* to obtain the t-score, p-value, and FDR for the groups of 152 normal uterine samples and 63 samples with uterine cervical cancer. Finally, we chose those genes having FDR$<$ 0.001 as differentially expressed. All of the differentially expressed genes were nevertheless taken into account as a single potential gene signature that could be validated during classification analysis by deep learning.

2.2.3 Disease classification of DEGs through deep learning

Here, we employed the newest deep learning technique,*nnet* (feed-forward neural network based model) [21]. We examined the class-label (normal and uterine cervical cancer groups) of the differentially expressed genes with a repeat of 30 times using this deep learning technique with 10-fold cross-validation. We segmented the predicted gene expression data for the DEGs into 10 folds of samples for the cross-validation, of which 9 folds served as the training set and the remaining fold served as the test set. As a result of this sub-step, we used the *nnet* tool with a specific number of epochs (referred to as "maxit"), which indicates that the deep learning process was internally repeated for that "maxit" times, and we computed the classification metrics after one iteration of each fold. A confusion matrix composed of true positive (TP), false negative (FN), false positive (FP), and true negative (TN) was obtained from this sub-step. For each fold of samples, same sub-process is repeated (i.e., other nine-fold samples). Then, we combined all of these parameters for the internal repeat that occurred 10 times, and a final confusion matrix was created. In order to acquire the average classification metric values (average accuracy, average sensitivity, average specificity, average precision, average overall error rate, and area under curve (AUC)), we then repeated the entire process a total of 30 times. In this case, we also applied a test sample as a validation sample. In this deep learning technique, we used *nnet* with the default parameter settings (i) size (= number of units in hidden layer), (ii) rang (= initial random weights) while [-rang, rang], (iii) decay (= parameter for weight decay), (iv) maxit (= the maximum number of iterations or number of epochs), (v) MaxNWts (= the maximum allowable number of weights), and other default parameters as well.

2.2.4 Hub gene finding

In order to determine the active edges among genes with correlation values ≥ 0.8 or ≤ -0.8, we conducted Pearson's correlation analysis to the DEGs found by our technique. We used the *Cytoscape* online tool [14] to perform degree centrality analysis after acquiring the collection of active edges, calculating the in-degree and out-degree scores of each DEG. In addition, we noted the top 10 in-degree hub DEGs and the top 10 out-degree hub DEGs.

2.2.5 Gene set enrichment analysis

Gene set enrichment analysis (GSEA) can assess the potential function, biological impact, and disease implications of a set of signature genes. After finding differentially expressed genes, we analyzed a list of the top differentially expressed genes by *WebGestalt (WEB-based Gene SeT AnaLysis Toolkit)* [20] using KEGG pathways and gene ontology (GO) annotations (three domains: biological process (BP), cellular component (CC), and molecular function (MF)). The number of genes included in each KEGG pathway and gene ontology (GO) term, as well as the enriched p-value and FDR, were obtained. When the FDR was greater than or equal to 0.05, we omitted those KEGG pathways or GO keywords from consideration.

2.3 Results and Discussion

In this case study, there are 215 samples, comprising 152 normal samples and 63 samples with uterine cervical cancer, and there are 27,578 characteristics. We retrieved 6,287 DEGs with FDR$<$ 0.001 by *Limma* after data preprocessing, linear regression, and differential expression analysis, in a list with estimated t-score, p-value, and FDR. Top 25 DEGs are listed in Table 2.1. For instance, *ADCY2* is the top DEG with the lowest FDR (FDR = 5.64×10^{-119}). In a supplementary file, we presented a list of all DEGs identified through differential expression analysis by empirical Bayes test using *Limma* with FDR corrected p-value. Additionally, another supplementary file was also available, which contained the predicted gene expression matrix of all DEGs computed from the original pre-filtered uterine cervical cancer DNA methylation data by linear regression analysis.

Then, using the most recent deep learning technique, *nnet* (feed-forward neural network based model) [21], we applied it to our obtained DEG expression dataset, which has 215 samples and 6287 attributes. We examined the class-label (normal and uterine cervical cancer groups) of the differentially expressed genes with a repeat of 30 times using this deep learning technique with 10-fold cross-validation. All of the samples from the predicted

Table 2.1 List of differentially expressed genes (FDR sorted).

Gene symbol	t-score	P-value	FDR
ADCY2	45.22	5.64×10^{-119}	5.95×10^{-115}
PTPN6	32.50	1.43×10^{-89}	7.55×10^{-86}
LHFPL2	32.09	1.63×10^{-88}	5.71×10^{-85}
VAV1	30.24	1.41×10^{-83}	3.72×10^{-80}
EYA4	−29.40	2.97×10^{-81}	6.27×10^{-78}
PNPLA2	29.02	3.38×10^{-80}	5.94×10^{-77}
ARID3A	−28.71	2.37×10^{-79}	3.56×10^{-76}
HOXD10	28.19	6.97×10^{-78}	9.20×10^{-75}
TWIST1	27.69	1.85×10^{-76}	2.17×10^{-73}
BHMT	26.49	5.42×10^{-73}	5.72×10^{-70}
TSLP	26.25	2.76×10^{-72}	2.65×10^{-69}
ACCN4	26.16	5.23×10^{-72}	4.60×10^{-69}
HOXA6	25.94	2.22×10^{-71}	1.80×10^{-68}
PRR5	25.67	1.40×10^{-70}	1.06×10^{-67}
NODAL	25.45	6.44×10^{-70}	4.53×10^{-67}
EFCAB1	25.41	8.60×10^{-70}	5.45×10^{-67}
WNT2	25.40	8.86×10^{-70}	5.45×10^{-67}
PC	25.40	9.31×10^{-70}	5.45×10^{-67}
S100A8	25.23	2.84×10^{-69}	1.58×10^{-66}
VWCE	24.89	3.05×10^{-68}	1.61×10^{-65}
IGFBP2	24.86	3.61×10^{-68}	1.81×10^{-65}
ZNF385A	24.74	8.36×10^{-68}	4.01×10^{-65}
C1orf220	24.71	1.04×10^{-67}	4.75×10^{-65}
COG2	24.63	1.85×10^{-67}	8.15×10^{-65}
QRFP	−24.51	4.16×10^{-67}	1.76×10^{-64}

gene expression data from the DEGs were split into 10 folds for the cross-validation, of which 9 folds served as the training set and the remaining fold served as the test set. In the subsequent sub-step, we used the *nnet* tool with a maxit (number of epochs) of 100, which meant that the deep learning method was internally repeated 100 times. We then computed the classification metrics in a single iteration of each fold. A confusion matrix made up of true positive (TP), false negative (FN), false positive (FP), and true negative (TN) was obtained from this sub-step. Each fold of samples undergoes this sub-process once (i.e., more nine other folds). The final confusion matrix was created after we combined all of these parameters for the 10 internal repetitions. After that, we carried out the entire process a total of 30 times to acquire 30 confusion metrics. With the help of this, we were able to calculate the average values for the classification metrics (average accuracy, average sensitivity, average specificity, average precision, average overall error rate, and area under curve (AUC)). Note that our deep learning approach has previously repeated 30,000 times ($30 \times 10 \times 100$) from which we derived average accuracy, where every sample has been utilized as test sample at least once (i.e., no sample is missing as test sample). Here we used the test sample as validation sample. In this deep learning method, we used $nnet$ with the default parameter settings (i) size (= number of units in hidden layer) (= 2), (ii) rang (= initial random weights)(= 0.1) while [-rang, rang], (iii) decay (= parameter for weight decay)(= $5 \times e^{-4}$), (iv) maxit (= the maximum number of iterations or number of epochs)(= 100), (v) MaxNWts (= the maximum allowable number of weights)(= 84,581), and other default parameters. As we employed 10-fold cross-validation, 9/10 of the 215 samples, or 194 or 193 samples, were taken as the training set, while 1/10 of the 215 samples, or 21 or 22, were taken as the test set. The remaining one-fold of samples served as the test set, with the remaining nine-fold of samples serving as the training set. Each sample must thus play each role at least once, whether in the training sample or the test sample. In this case, we also employed a test sample as a validation sample. Our average classification accuracy was 90.69% ($\pm 1.97\%$), and the AUC value was 0.858. For more details, see Table 2.2. We have plotted all metrics in Figure 2.2.

We conducted a comparison between the proposed technique and an existing method called $RSNNS$ (a deep learning tool based on the Stuttgart Neural Network Simulator (SNNS)), using 10-fold cross-validation and repeating the experiment 30 times. In the case of $RSNNS$, we also used the same default parameter settings such as (i) size (= number of units in hidden layer) (= 2), and (ii) maxit (= maximum number of iterations or number of epochs) (= 100), among others. We ran through the entire process in both instances 30 times to ensure a fair classification. In contrast to the existing approach $RSNNS$, which had an average classification accuracy of 87.27% ($\pm 5.92\%$), our new method produced an average classification accuracy of 90.69% ($\pm 1.97\%$) (see Figure 2.3). We believed that our framework performed better than any previous deep learning methods.

Table 2.2 Values of disease classification metrics by the proposed method.

Metrics	Average value (std)
Average accuracy	90.69% ($\pm 1.97\%$)
Average sensitivity	73.97% ($\pm 1.06\%$)
Average specificity	97.63% ($\pm 1.71\%$)
Average precision	93.38% ($\pm 4.17\%$)
Average overall error rate	9.30% ($\pm 1.97\%$)
Area under curve (AUC)	0.858

$*$std: standard deviation.

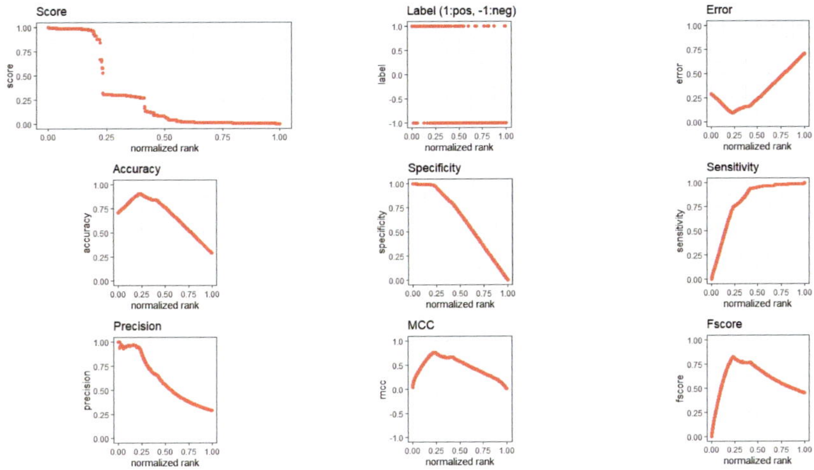

Figure 2.2 ROC plots of all classification metrics for the proposed method.

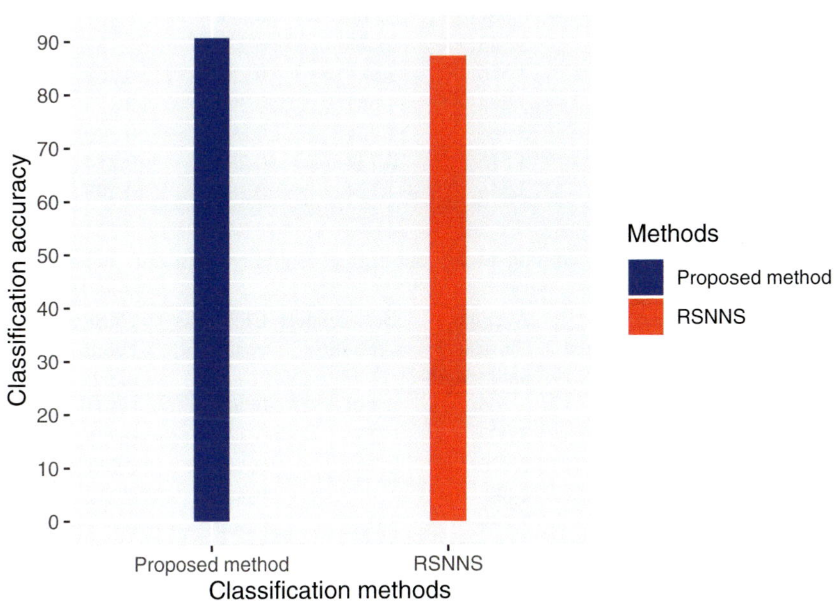

Figure 2.3 Comparative bar plot: proposed method vs. state-of-the-art method (RSNNS).

In order to identify edges between genes with correlation values larger than or equal to 0.8 or less than or equal to (-0.8), we used Pearson's correlation analysis on our DEGs. Subsequently, utilizing *Cytoscape* [14], we also carried out in-degree and out-degree hub gene network analyses. As an illustration, the top 5 genes with the greatest in-degree values are *PAIP2*, *GRWD1*, *VPS4B*, *CRADD*, and *LLPH*, as shown in Table 2.3. The top 5 most out-degree genes are also listed in Table 2.4, and they are *MRPL35*, *FAM177A1*, *STAT4*, *ASPSCR1*, and *FABP7*. We detailed the hub gene network structure in a supplementary file.

In the associated literature review, we discovered that the majority of the top hub genes identified by our technique had a significant impact

Table 2.3 Top 10 hub genes according to the in-degree centrality from our proposed method.

Gene symbol	In-degree	Out-degree	Average shortest path length	Betweenness centrality	Closeness centrality	Clustering coeffi-cient
PAIP2	439	32	3.587	0.802	0.279	0.188
GRWD1	425	66	3.435	11.001	0.291	0.178
VPS4B	406	68	3.460	2.276	0.289	0.191
CRADD	406	178	3.087	11.003	0.324	0.152
LLPH	403	40	3.545	2.313	0.282	0.182
NDUFA4	390	89	3.556	1.927	0.281	0.168
NDUFB6	372	111	3.294	4.661	0.304	0.175
ZKSCAN4	372	88	3.364	1.434	0.297	0.200
SMARCD1	365	43	3.515	0.734	0.284	0.214
TMED10	348	39	3.546	4.124	0.282	0.193

Table 2.4 Top 10 hub genes according to the out-degree centrality from our proposed method.

Gene symbol	In-degree	Out-degree	Average shortest path length	Betweenness centrality	Closeness centrality	Clustering coeffi-cient
MRPL35	239	376	2.765	9.354	0.362	0.141
FAM177A1	21	339	3.002	0.263	0.333	0.225
STAT4	94	332	2.872	2.744	0.348	0.211
ASPSCR1	68	329	2.888	1.132	0.346	0.212
FABP7	204	315	2.779	3.008	0.360	0.171
HNRNPA0	65	311	3.010	1.230	0.332	0.191
ANGPTL4	18	299	2.887	0.364	0.346	0.249
DDX19A	86	283	2.993	1.385	0.334	0.218
TRNT1	40	282	3.119	0.477	0.321	0.221
PFDN1	52	274	3.005	0.526	0.333	0.243

on the particular cancer. According to Berlanga et al. [23], the *PAIP2* gene and cervical cancer are related. *GRWD1* was used to control p53 adversely throughout the development of tumors [24]. Moreover, it had been explored as a potential biomarker for DNA methylation during cancer treatment and risk assessment. The methylation of *GRWD1* may operate as a preventative measure in the growth of tumors [25]. In the literature by Broniarczyk et al. [26], there are reports of the *VPS4B* gene with cervical cancer. Similar to how the *LLPH* gene was linked to cervical cancer, the *CRADD* gene is also connected in this disease, according to Feron et al. [28] and Sundaram et al. [27] respectively. Similar to this, a literature search revealed that the top out-degree hub genes were primarily linked to cervical cancer. For instance, the link between *STAT4* and the respective cervical cancer was shown in Luo et al. [30], whilst the association between *FAM177A1* and cervical cancer was recorded in [29]. Also, Liang et al. [31] produced an evidence on *ASPSCR1* and cervical cancer, and Zhang et al. [32] discovered a connection between *FABP7* and cervical cancer.

These 6287 DEGs with FDR< 0.001 are used for gene set enrichment analysis by *WebGestalt (WEB-based Gene SeT AnaLysis Toolkit)* [20]. In order to get all KEGG pathways and gene ontology (GO) terms (biological process (BP), cellular component (CC), and molecular function (MF)), along with the number of genes in each pathway or GO-term, enriched *p*-value, and FDR, we applied *WebGestalt* database to our DEG set. Here, we used the input dataset that *WebGestalt* specifies, which is a two-column pattern with the first column representing the gene name and the second column

Table 2.5 Top significant KEGG pathways (FDR sorted).

KEGG pathway name	#genes	Enriched *p*-value	FDR
hsa05205 Proteoglycans in cancer	198	6.65×10^{-08}	2.16×10^{-05}
hsa04550 Signaling pathways regulating pluripotency of stem cells	139	1.32×10^{-07}	2.16×10^{-05}
hsa05166 Human T-cell leukemia virus 1 infection	255	6.95×10^{-07}	7.29×10^{-05}
hsa04510 Focal adhesion	199	8.94×10^{-07}	7.29×10^{-05}
hsa05200 Pathways in cancer	524	1.26×10^{-06}	8.19×10^{-05}
hsa04015 Rap1 signaling pathway	206	1.93×10^{-06}	1.05×10^{-04}
hsa04514 Cell adhesion molecules (CAMs)	144	2.31×10^{-07}	1.07×10^{-04}
hsa04611 Platelet activation	123	7.22×10^{-06}	2.94×10^{-04}
hsa04072 Phospholipase D signaling pathway	146	1.02×10^{-05}	3.69×10^{-04}
hsa04640 Hematopoietic cell lineage	97	4.01×10^{-05}	1.31×10^{-03}

representing the score. Here, the *t*-score was used as the score. Significant pathways and GO-terms are discussed, along with Table 2.5–Table 2.8. For example, the KEGG pathway *hsa05205:Proteoglycans in cancer* was the most significant and had the lowest FDR value (2.16×10^{-05}). In total, 198 genes were linked to this pathway with an enriched *p*-value of 6.65×10^{-08}. See Table 2.5 for the remaining top 10 significant KEGG pathways. We included a supplementary file that contained a list of all KEGG pathways. Also, Figure 2.4 displays the volcano plot of the normalized enrichment

Table 2.6 Top significant GO-BP terms enriched (FDR sorted).

GO-BP term name	#genes	Enriched *p*-value	FDR
GO:0008283 cell proliferation	1986	0	0
GO:0006928 movement of cell or subcellular component	1967	0	0
GO:0009891 positive regulation of biosynthetic process	1949	0	0
GO:0016192 vesicle-mediated transport	1942	0	0
GO:0006955 immune response	1919	0	0
GO:0031328 positive regulation of cellular biosynthetic process	1919	0	0
GO:0006915 apoptotic process	1911	0	0
GO:0010628 positive regulation of gene expression	1911	0	0
GO:2000026 regulation of multicellular organismal development	1908	0	0
GO:0006468 protein phosphorylation	1860	0	0

Table 2.7 Top significant GO-CC terms enriched (FDR sorted).

GO-CC term name	#genes	Enriched *p*-value	FDR
GO:0005783 endoplasmic reticulum	1861	0	0
GO:0097458 neuron part	1690	0	0
GO:0031984 intrinsic component of plasma membrane	1673	0	0
GO:0031984 organelle subcompartment	1661	0	0
GO:0098805 whole membrane	1630	0	0
GO:0005887 integral component of plasma membrane	1596	0	0
GO:0005794 Golgi apparatus	1516	0	0
GO:0044433 cytoplasmic vesicle part	1462	0	0
GO:0044463 cell projection part	1425	0	0
GO:0120038 plasma membrane bounded cell projection part	1425	0	0

Table 2.8 Top significant GO-MF terms enriched (FDR sorted).

GO-MF term name	#genes	Enriched *p*-value	FDR
GO:0042802 identical protein binding	1696	0	0
GO:0005102 signaling receptor binding	1538	0	0
GO:0019904 protein domain specific binding	684	0	0
GO:0044212 transcription regulatory region DNA binding	896	1.33×10^{-15}	6.25×10^{-13}
GO:0001067 regulatory region nucleic acid binding	898	2.00×10^{-15}	7.50×10^{-13}
GO:0003690 double-stranded DNA binding	915	1.02×10^{-14}	3.16×10^{-12}
GO:0008134 transcription factor binding	638	1.18×10^{-14}	3.16×10^{-12}
GO:0016301 kinase activity	845	2.19×10^{-14}	5.00×10^{-12}
GO:1990837 sequence-specific double-stranded DNA binding	823	2.49×10^{-14}	5.00×10^{-12}
GO:0000976 transcription regulatory region sequence-specific DNA binding	781	2.66×10^{-14}	5.00×10^{-12}

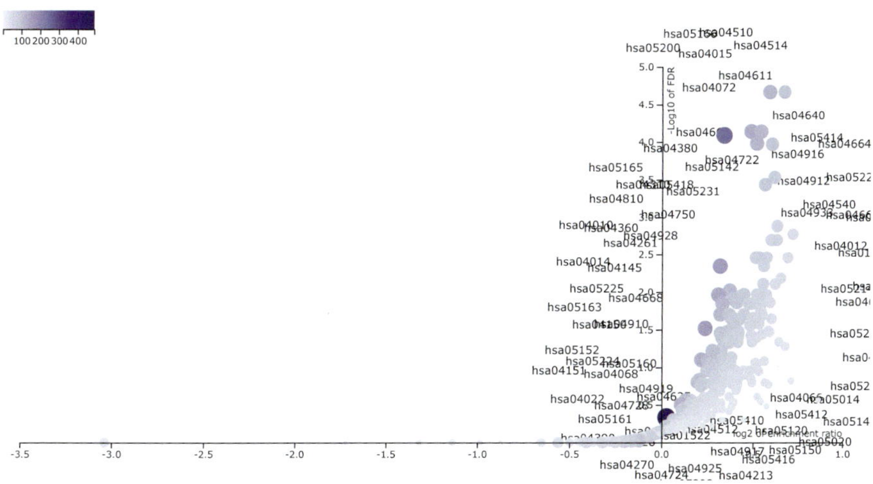

Figure 2.4 The volcano plot of normalized enrichment score of the FDR significant KEGG pathways from GSEA analysis of DEGs.

score for those KEGG pathways that are FDR significant. The top significant GO-BP term with an FDR value of zero was *GO:0008283 cell proliferation*. With an enriched *p*-value of 0, a total of 1986 genes were connected to this GO-BP term. See Table 2.6 for the remaining terms. As a supplementary file, we presented the whole list of GO-BP terms. *GO:0005783 endoplasmic reticulum* was identified as one of the most significant GO-CC terms in our analysis, having an FDR value of 0. With an enriched p-value of 0, a total of 1861 genes were connected to this GO-CC term. See Table 2.7 for the rest. We provided the list of all GO-CC terms in a supplementary file. Additionally, one of the most significant GO-MF terms with a minimal FDR value of 0 was *GO:0042802 identical protein binding*. This GO-MF term was linked to 1696 different genes with an enriched p-value 0. See Table 2.8 for more information. In a supplementary file, we included a list of all GO-MF terms.

2.4 Conclusion and Future Work

In this chapter, we presented a methodology for integrating DNA methylation and associated TCGA gene expression data to uterine cervical cancer utilizing linear regression, differential expression, and deep learning. Prior to applying linear regression to estimate predicted gene expression data from the preprocessed DNA methylation data using TCGA gene expression data, we first deleted outliers. Then, employing downstream analysis and the empirical Bayes test with $Limma$, we found 6287 genes that were differentially expressed with an FDR cutoff of less than 0.001. After that, we used the $nnet$ deep learning approach to evaluate differentially expressed genes with 10-fold cross-validation and the default parameter settings (i) size (= number of units in hidden layer), (ii) rang (= initial random weights) while [-rang, rang], (iii) decay (= parameter for weight decay), (iv) maxit (= the maximum number of iterations or number of epochs), (v) MaxNWts (= the maximum allowable number of weights), as well as other default parameters also. Our average classification accuracy for the normal and uterine cervical cancer samples using DEG expression data was $90.69\%(\pm1.97\%)$, which is significantly higher than that of traditional works. We can conclude that our obtained DEGs are powerful and effective in classifying diseases as a result of deep learning and comparison analysis.

Using $Cytoscape$ online tool (version 3.8.0), we also carried both in-degree and out-degree hub gene network analyse in this case. The five genes with the highest in-degree expression were $PAIP2$, $GRWD1$, $VPS4B$, $CRADD$, and $LLPH$, while the five genes with the highest out-degree expression were $MRPL35$, $FAM177A1$, $STAT4$, $ASPSCR1$, and $FABP7$. Additionally, using $WebGestalt$, we performed pathway analysis on DEGs with $FDR < 0.001$. Finally, our framework is more beneficial for improving biological interpretation of the DNA methylation data than just performing a single differential methylation study or identifying differentially methylated regions.

In future study, we may build on our existing work by fusing the random forest ensemble method with deep learning strategy to produce a better classification model from every aspect, and then apply that to big data (such as single-cell RNA sequencing data or other data specifically related to TCGA cancer tissue) for the classification of cancer.

References

[1] B. Jin, Y. Li, and K. D. Robertson, DNA Methylation: Superior or Subordinate in the Epigenetic Hierarchy?, Genes & Cancer, vol 2 no 6 pp 607–617, 2011, DOI: 10.1177/1947601910393957

[2] Kim,J.H. et al. (2012) LRpath analysis reveals common pathways dysregulated via DNA methylation across cancer types. BMC Genomics, 13, 526.

[3] S. Mallik, A. Mukhopadhyay and U. Maulik, Integrated Statistical and Rule-Mining Techniques for DNA Methylation and Gene Expression Data Analysis, JAISCR, 2013, Vol. 3, No. 2, pp. 101-115, DOI 10.2478/jaiscr-2014-0008

[4] Marsit,C.J. et al. (2011) DNA methylation array analysis identifies profiles of blood-derived DNA methylation associated with bladder cancer. J. Clin. Oncol., 29, 1133–1139.

[5] Rijlaarsdam,M.A. et al. (2014) DMRforPairs: identifying differentially methylated regions between unique samples using array based methylation profiles, BMC Bioinformatics, 15, 141.

[6] S. Mallik, G. J. Odom, Z. Gao, L. Gomez, X. Chen, L. Wang, An evaluation of supervised methods for identifying differentially methylated regions in Illumina methylation arrays, Briefings in Bioinformatics, Volume 20, Issue 6, November 2019, Pages 2224–2235, https://doi.org/10.1093/bib/bby085.

[7] Y. Wang, J. M. Franks, M. L. Whitfield and C. Cheng, BioMethyl: an R package for biological interpretation of DNA methylation data, Bioinformatics, 35(19), 2019, 3635–3641 doi: 10.1093/bioinformatics/btz137.

[8] McLean,C.Y. et al. (2010) GREAT improves functional interpretation of cis-regulatory regions. Nat. Biotechnol., 28, 495–501.

[9] Goecks,J. et al. (2010) Galaxy: a comprehensive approach for supporting accessible, reproducible, and transparent computational research in the life sciences. Genome Biol., 11, R86.

[10] https://www.healthline.com/health/cervical-cancer

[11] Zhuang J, Jones A, Lee SH, Ng E et al. The dynamics and prognostic potential of DNA methylation changes at stem cell gene loci in women's cancer. PLoS Genet 2012 Feb;8(2):e1002517. PMID: 22346766

[12] Teschendorff AE, Jones A, Fiegl H, Sargent A et al. Epigenetic variability in cells of normal cytology is associated with the risk of future morphological transformation. Genome Med 2012 Mar 27;4(3):24. PMID: 22453031

[13] Teschendorff AE, Jones A, Widschwendter M. Stochastic epigenetic outliers can define field defects in cancer. BMC Bioinformatics 2016 Apr 22;17:178. PMID: 27103033

[14] Ritchie ME, Phipson B, Wu D, et al. limma powers differential expression analyses for RNA-sequencing and microarray studies. Nucleic Acids Res. 2015; 43(7):e47.

[15] Xu,Z. et al. (2016), ENmix: a novel background correction method for Illumina HumanMethylation450 BeadChip. Nucleic Acids Res., 44, e20.

[16] S. Bandyopadhyay, S. Mallik, A. Mukhopadhyay, A survey and comparative study of statistical tests for identifying differential expression from microarray data. IEEE/ACM Trans. Comput. Biol. Bioinf. 11(1), 95–115 (2014).

[17] C. W. Law, Y. Chen, W. Shi and G. K. Smyth, voom: precision weights unlock linear model analysis tools for RNA-seq read counts, Genome Biol 2014, 15:R29.

[18] S. Mallik, S. Seth, T. Bhadra, N. Tomar, Z. Zhao, A Multi-classifier Model to Identify Mitochondrial Respiratory Gene Signatures in Human Cancer, 2019 IEEE International Conference on Bioinformatics and Biomedicine (BIBM), ISBN: 978-1-7281-1867-3, pp. 1928-1935.

[19] S. Mallik, A. Mukhopadhyay, and U. Maulik, RANWAR: Rank-Based Weighted Association Rule Mining From Gene Expression and Methylation Data, IEEE TRANS ON NANOBIOSCIENCE, VOL. 14, NO. 1, pp. 59–66, JANUARY 2015.

[20] Y. Liao, J. Wang, E. J. Jaehnig, Z. Shi, B. Zhang, WebGestalt 2019: gene set analysis toolkit with revamped UIs and APIs, Nucleic Acids Research, Volume 47, Issue W1, 02 July 2019, Pages W199–W205, https://doi.org/10.1093/nar/gkz401.

[21] W. N. Venables and B. D. Ripley, Feed-Forward Neural Networks and Multinomial Log-Linear Models (Package "nnet"), CRAN R project repository, 2020.

[22] P. Shannon, A. Markiel, O. Ozier, N. S. Baliga, J. T. Wang, D. Ramage, N. Amin, B. Schwikowski, T. Ideker, Cytoscape: A Software Environment for Integrated Models of Biomolecular Interaction Networks, Genome Res. 2003 Nov; 13(11): 2498-2504. doi: 10.1101/gr.1239303. PMCID: PMC403769

[23] J. J. Berlanga1, A. Baass1, and N. Sonenberg, Regulation of poly(A) binding protein function in translation: Characterization of the Paip2 homolog, Paip2B, RNA (2006), 12:1556-1568. Published by Cold Spring Harbor Laboratory Press.

[24] K. Kayama, S. Watanabe, T. Takafuji, T. Tsuji, K. Hironaka1, M. Matsumoto, K. I. Nakayama, M. Enari, T. Kohno, K. Shiraishi, T. Kiyono, K. Yoshida1, N. Sugimoto1 and M. Fujita1, GRWD1 negatively regulates p53 via the RPL11-MDM2 pathway and promotes tumorigenesis, EMBO reports, Vol. 18, No. 1, 2017.

[25] C. Gao, J. Zhuang, H. Li1, C. Liu, C. Zhou, L. Liu and C. Sun, Exploration of methylation-driven genes for monitoring and prognosis of patients with lung adenocarcinoma, Cancer Cell International(2018), 18:194, https://doi.org/10.1186/s12935-018-0691-z.

[26] J. Broniarczyk, D. Pim, P. Massimi, M. Bergant, A. Jozefiak, C. Crump and L. Banks, The VPS4 component of the ESCRT machinery plays an essential role in HPV infectious entry and capsid disassembly, Scientific Reports, 7:45159, DOI: 10.1038/srep45159.

[27] M. K. Sundaram, R. Raina, N. Afroze, K. Bajbouj, M. Hamad, S. Haque and A. Hussain, Quercetin modulates signaling pathways and induces apoptosis in cervical cancer cells, Bioscience Reports (2019) 39 BSR20190720, https://doi.org/10.1042/BSR20190720

[28] O. Feron, R. Boidot, S. Branders, P. Dupont, T. Helleputte, Signature of cycling hypoxia and use thereof for the prognosis of cancer, International Application Published under the Patent Cooperation Treaty (PCT), WO 2015/015000 Al, https://patentimages.storage.googleapis.com/80/1a/3c/eac6d250b2943a/WO2015015000A1.pdf

[29] H. Wen, L. Chen, J. He and J. Lin, MicroRNA expression profiles and networks in placentas complicated with selective intrauterine growth restriction, Molecular Medicine Reports 16: 6650-6673, 2017, DOI: 10.3892/mmr.2017.7462.

[30] J. Luo, Q. Huang, X. Lin, STAT4 expression is correlated with clinicopathological characteristics of cervical lesions, https://www.researchgate.net/publication/303787920_STAT4_expression_is_correlated_with_clinicopathological_characteristics_of_cervical_lesions/citations

[31] W. S. Liang, J. Aldrich, S. Nasser, Simultaneous Characterization of Somatic Events and HPV-18 Integration in a Metastatic Cervical Carcinoma Patient Using DNA and RNA Sequencing, International Journal of Gynecological Cancer & Volume 24, Number 2, February 2014.

[32] C. Zhang, Y. Liao, P. Liu, Q. Du, Y. Liang, S. Ooi, S. Qin, S. He, S. Yao, W. Wang, FABP5 promotes lymph node metastasis in cervical cancer by reprogramming fatty acid metabolism, Theranostics 2020, Vol. 10, Issue 15, pp. 6561-6580. doi: 10.7150/thno.44868.

[33] S. Mallik, Z. Zhao, "Graph- and rule-based learning algorithms: A comprehensive review of their applications for cancer type classification and prognosis using genomic data", Briefings in Bioinformatics, bby120, 2018.

[34] G. Qin, S. Mallik, R. Mitra, A. Li, P. Jia, C. Eischen, Z. Zhao, "MicroRNA and transcription factor co-regulatory networks and subtype classification of seminoma and non-seminoma in testicular germ cell tumors", Nature Scientific Reports, 2019,

[35] S. Mallik, G. Qin, P. Jia, Z. Zhao, "Molecular signatures identified by integrating gene expression and methylation in non-seminoma and seminoma of testicular germ cell tumors", Epigenetics, Taylor & Francis, 2020

[36] Mallik S, Seth S, Bhadra T, Zhao Z. A Linear Regression and Deep Learning Approach for Detecting Reliable Genetic Alterations in Cancer Using DNA Methylation and Gene Expression Data, Genes (Basel), 2020 Aug 12; 11(8):931. doi: 10.3390/genes11080931. PMID: 32806782, PMCID: PMC7465138.

3

Multi-objective Optimization-based Association Rule Mining Integral Approach for Optimal Ranking and Directional Signature Classification of Multi-omics Data

3.1 Introduction

Microarray [1], [2] has been frequently utilized in gene expression profile, methylation, and genotype-based association studies to test a large number of genes for changes between two groups (for example, cases versus control samples). Without altering the underlying DNA sequences, methylation of cytosine [3] modifies the DNA structure by adding a methyl group (-CH3) at the position of carbon5 of the cytosine. Methylation modifies gene expression, which inevitably causes cancer. Methylation often lowers the degree of gene expression. In the realm of bioinformatics and biological research, the association rule mining (ARM) approach is currently essential for producing the significant relationships between two genes (items) [4]. The rule is written in the following format: $\{G1+, G2-, G4+ \Rightarrow G3-, G5+\}$, where $G1$, $G2$, and $G4$ are the cause variables (antecedent) and $G3$ and $G5$ are the impacted variables (consequent). Notably, the "+" here represents up-regulation while the "-" represents down-regulation. According to the aforementioned rule, it is likely that $G3$ will be down-regulated and $G5$ will be up-regulated simultaneously when the genes $G1$ is up-regulated, $G2$ is down-regulated, and $G4$ is up-regulated concurrently. To define the support of a rule $\{A \Rightarrow B\}$ (A and B are items) here used the ratio of the number of transactions that contains A and B and the total number of transactions in the database. Alternatively, the confidence of a rule is defined by dividing the support of the entire gene set (i.e., A and B) by the support of the antecedent/left-hand side (i.e., A). The geneset is referred to as frequent if its support is more than the minimum support specified by the user. Agrawal et al. [6] introduced

Apriori, an efficient and essential association rule mining method, in 1993 for determining the association among the genes in the gene expression data or other similar types of data. Another benchmark algorithm for frequent itemset mining is Eclat (Equivalence class clustering and bottom-up lattice traversal), which was developed by Zaki in 2000 [7]. The fundamental steps of the Apriori algorithm are as follows: i) determining the support (frequency) value for each individual item (feature); ii) removing non-frequent items using a user-defined support threshold; iii) choosing frequent k-item sets (k = 1,2,...); iv) converting all frequent item sets into association rules; and v) finally, estimating two additional rule-interestingness measures, namely confidence and lift. The Eclat algorithm [7], however, differs somewhat from the Apriori algorithm [6]. Eclat is a tree-based algorithm, whereas Apriori is essentially a join-based algorithm. In other words, Apriori uses breadth-first search, i.e., horizontal search, whereas Eclat uses depth-first search, i.e., vertical search. Faster than Apriori is Eclat. For the Eclat algorithm, only the support metric is required. Both algorithms employ static confidence and support thresholds.

Researchers like Han et al. [12], Creighton and Hanash [9], Georgii et al. [10], McIntosh and Chawla [11], Martinez et al. [13], and others have updated and altered this fundamental ARM approach based on the problem types to solve the different challenges. These modern methods assist us in managing the significant issues that crop up in our day-to-day lives, such as travel, marketing, and medical diagnosis. Different forms of priority are assigned to the genes in the genesets. However, traditional rule mining algorithms treat every gene in the geneset as being of equal class (quality). To address this issue, the researchers listed below developed weighted ARM methods for classifying genes (viz., Ramkumar et al. [15], Cai et al. [17], Wang et al. [18], Tao et al. [21], Yun and Leggett [20], Tseng et al. [22], and $RANWAR$ [14]). To address a problem involving transaction data, the weighted ARM approaches were improved further and taken into account as multiple weighted components (Liu et al. [44], Su et al. [43], and Liu et al. [19]). For the purpose of analyzing gene expression data, certain clustering and biclustering-based algorithms are developed by Cheng and Church [24], Pei et al. [25], Jiang et al. [26], Madeira and Oliveira [23], Thalamuthu et al. [27], Prelic et al. [28], and StatBicRM [16]. For this reason, another classification analysis was also devised (Bhasin and Raghava [29], Martella [31], and Liu and Xu [32]). In order to obtain quantitative association rules from numeric microarray datasets without applying the discretization, Georgii et al. (in 2005) [10] employed a half-space notion. The disadvantage of this method was that it was unable to extract the entire set of significant rules from the microarray data. Martinez et al. (in 2008) [13] developed the GenMiner method for determining association rules using a set of gene expression data and the terms that were easily accessible online and related to gene ontology (i.e., GO-terms). In a novel approach provided by Bhadra et al. [39], the cutoff (threshold) value was dynamically taken into account and changed for each geneset in accordance with the quality or significance of the entire geneset rather than a quantitative property. There are also some recent works that use association rule mining (ARM)/optimization. A tool was offered for two-arm clinical research by Theilhaber et al. [49]. The methodology was based on the development and improvement of a multivariate predictive gene signature that can predict the varied survival of patients receiving anti-cancer treatments [49]. Murphy et al. [50] used enhanced binary particle swarm optimization (EBPSO) in clinical transcriptomics cohorts to find precise, distinct candidate signatures that were significantly predictive. The gene regulator contained in this signature provides biological insights into the pertinent functions that were closely associated with their form of cancer [50]. Different feature selection techniques were categorized by Mahendran et al. [51] into supervised, unsupervised, and semi-supervised learning in order to provide a research study. This study presented some of the most recent dimensionality reduction techniques for cancer identification, evaluated their efficacy, indicated their shortcomings, and suggested future research directions for handling high-dimensional data and small sample sizes. A modeling, simulation, and performance analysis study for an isolated hybrid power system (IHPS) that included a solar thermal power plant, a diesel engine generator (DEG), and a wind turbine generator (WTG) was presented by Ganguly et al. in 2020 [52]. The authors used a quasi-oppositional-based whale optimization technique to enhance the results for the IHPS model under study and outperformed other benchmarks in terms of controller gain. Additionally, there are several existing rule mining works that are based on fuzzy or rough theory [55], [56]. The outcome rules, however, need improvement. An effective way to boost the performance of association rule mining is to incorporate multi-objective optimization techniques. We obtained some recently developed multi-objective optimization strategies after reviewing the literature, which are provided by Mudi et al. [57], Mudi et al. [58], [59], Mudi et al. [60], Ganguly et al. [61], [62], Ganguly and Mukherjee [63], and so on. In this chapter, we present the development of multi-objective optimized variable cutoff-based association rule mining ($MOOVARM$) for multi-omics profile based on minimum distance from positive ideal solution (PIS) and minimum distance from negative ideal solution (NIS). In order to address this, we first determined the positive ideal solution (PIS), and negative ideal solution (NIS) in regard to all genesets. As a result, we determined the distances ($d+$) from PIS and ($d-$) from NIS for each product or item set. We determined the relative closeness score value for ranking the genesets using our proposed approach based on the two distances, $d+$ and $d-$. Any rule might be regarded as a final resultant rule if its relative closeness score exceeded and was equal to the predetermined cutoff value. Instead of using individual genes, the proposed method determined the global relative closeness score. Last but not least, we created a ranking of the rules based on the relative score that can be used to provide improved disease classification performance compared to the most state-of-the-art algorithms in disease diagnosis and therapeutic response.

3.2 Shortest Distance-based Cutoffs

The distance-based variable support (abbreviated D_bVS) pruning technique, introduced by Mallik and Zhao [36], was adopted to obtain attractive rules from multi-omics datasets by combining co-expression, co-methylation, and protein–protein interactions. The normalized combination correlation score was determined by integrating the co-expression and co-methylation values (for example, $CECM_{exm}$) between the expression and methylation profiles while taking into account a certain set of genes that exhibit both differential expression and methylation. Essentially, $CECM_{exm}$ compares the methylation and expression patterns of the two genes. A gene vector G stands for the expression/methylation data for all diseased and control values. Let p and q be two genes. $CECM_{exm}(p, q)$ stands for the $CECM_{exm}$ between p and q. The calculation is as follows:

$$CECM_{exm}(p, q) = \text{norm}(\text{PCB}(G_{ex}(p), G_{ex}(q)) * r(G_m(p), G_m(q))), \tag{3.1}$$

where $G_{ex}(p)$ and $G_m(p)$ are two vectors that represent the expression and methylation levels for the pth gene, respectively, across all samples. The Pearson's correlation coefficient [4] between the two groups is symbolized by $r(\cdot, \cdot)$. The multiplication of the Pearson's correlation score and the BioSim score [5] between any two genes is performed by the PCB$(\cdot, \cdot)$. Here, the min–max normalization notion is used to indicate the normalization process using the symbol norm$(\cdot)$. The lower and upper bounds of $CECM_{exm}(\cdot, \cdot)$ were 0 and 1, respectively. Then, using the $CECM_{exm}$ scores, the corresponding dissimilarity scores (let us say, D_{simt}) were calculated, i.e.,

$D_{simt}(p,q) = (1 - CECM_{exm}(p,q))$. Afterwards, using the Human Protein Resource Database ($HPRD$), we identified protein–protein interactions and chose the interactions for which the interactive protein-oriented genes were expressed and methylated. The protein-protein interaction matrix for the chosen differentially expressed and methylated genes is represented by the symbol H. We generated the resultant value, say $DijStP(p,q)$, by multiplying the interaction value in H with the matching weighted distance value in D_{simt} for each gene pair (p,q). $DijStP(p,q) = (H(p,q) * D_{simt}(p,q))$ is the expression for $DijStP(p,q)$. We selected the weighted distance for each gene pair that includes the interactions in their corresponding protein levels in order to compute $DijStP(p,q)$ for all gene pairs (p,q). This outcome consisted of a symmetric as well as similarity matrix. We created a weighted transcriptomic gene regulatory network using this matrix. The gene regulatory network was then subjected to the Dijkstra shortest path algorithm, which produced the relative weighted shortest distance matrix (designated by W_eSD). The fundamental biological hypothesis states that if two genes are close enough to one another, their biological functions or biological pathways will be identical. In this study, we used the shortest distance between each pair of network-related genes. Then, we determined different distances between all gene pairs that were part of the W_eSD matrix, such as the maximum weighted shortest distance (W_eSD_{mx}), the minimum weighted shortest distance (W_eSD_{mn}), and the average weighted shortest distance (W_eSD_{avg}), which were all calculated without taking the diagonal elements of the underlying matrix into account. Within the geneset (GS), the distance-based variable supports threshold ($D_bVS(GS)$) is defined as follows:

$$D_bVS(GS) = \sqrt{\frac{1}{ngp} \sum_{p,q \in GS; p \neq q} WV_{msc}(p,q)^2}, \tag{3.2}$$

where

$$WV_{msc}(p,q) = \begin{cases} UVminS\left(1 - \frac{(WeSD(p,q) - med(WeSD)) * c1}{c2 * MAD(WeSD)}\right), & \text{if } p! = q \\ UVminS, & \text{if } p == q, \end{cases}$$

where ngp is the total number of gene pairs that could exist within GS. Two constant terms, $c1$ and $c2$, are present here. For $c1$, the value chosen is 0.10, and $c2$ is a constant scaling factor that is here taken to be 1.4826 for the purpose of utilizing any parametric test with the assumption of a Gaussian distribution pattern.

Similarly, the two additional threshold types are known as distance-oriented variable confidence (D_bVC) and distance-oriented variable lift (D_bVL), respectively. The following defines these two matrices:

$$D_bVC(GS) = \sqrt{\frac{1}{ngp} \sum_{p,q \in GS; p \neq q} WV_{mcc}(p,q)^2}, \tag{3.3}$$

where

$$WV_{mcc}(p,q) = \begin{cases} UVminC\left(1 - \frac{(WeSD(p,q) - med(WeSD)) * c1}{c2 * MAD(WeSD)}\right), & \text{if } p! = q \\ UVminC, & \text{if } p == q, \end{cases}$$

where $UVminC$ is the minimum confidence level specified by the user, and

$$D_bVL(GS) = \sqrt{\frac{1}{ngp} \sum_{p,q \in GS; p \neq q} WV_{mlc}(p,q)^2}, \tag{3.4}$$

where

$$WV_{mlc}(p,q) = \begin{cases} UDminL\left(1 - \frac{(WeSD(p,q) - med(WeSD)) * c1}{c2 * MAD(WeSD)}\right), & \text{if } p! = q \\ UDminL, & \text{if } p == q, \end{cases}$$

where $c1$ and $c2$ stand for the constant values for scaling the fractional component and $UDminL$ is the user-specified minimum lift threshold value.

3.3 Multi-objective Optimized Association Rule Mining for Multi-omics Dataset

In this section, we built an entirely novel method for multi-omics profiles called $MOOVARM$ (multi-objective optimized variable cutoff-based association rule mining) [70]. Here, we applied and integrated the gene expression, methylation, and protein–protein interaction data using the concepts of weighted shortest distance and multi-objective optimization to create interesting rules for multi-omics profiles. There are explanations for the first three steps of this method.

3.3.1 Finding significant genes

In the beginning, matched samples and genes between the methylation and gene expression data were discovered. The data on gene expression and methylation were normalized gene-wise using the zero-mean normalization technique [1]. For the goal of identifying differentially expressed/methylated genes, on the normalized expression/methylation data, an empirical Bayes test using Limma package [36], [35], [38] was used. Since Limma works with any data distribution and any number of samples, it was chosen because of its efficiency on normalized gene expression/methylation data. In the normalized expression/methylation dataset, many gene pairs contained multiple probes. We used Limma for each gene probe separately to identify the probes that were differentially expressed or methylated based on a significant Benjamini–Hochberg (BH) corrected p-value. The probes are classified as differentially expressed/methylated gene probes if the Benjamini–Hochberg (BH) corrected p-value is less than the accepted cutoff of 0.05 for the expression/methylation data. Then, we chose the probe of each gene for which the Benjamini–Hochberg (BH) corrected p-value produced by the Limma tool was to be lowest among all probes of each gene. From the matching dataset, the remaining probes for those genes were subtracted. Finally, only genes with single probes were found to be differentially expressed, methylated, and to have interactions amongst their corresponding proteins in the $HPRD$ database.

3.3.2 Discretization and post-discretization formats

Assume that n represents the number of genes that are both differentially expressed and differentially methylated (N) and that N represents the set of genes that had both the differential expression and differential methylation patterns and is involved in protein–protein interaction. And let m represent the quantity of samples that were matched between the expression and methylation data, whereas M represents the set of samples that were matched. DiE_{norm} and DiM_{norm}, respectively, stand for the normalized expression and methylation data matrices of the genes belonging to N. Gene was represented in the row of the data matrices, while the transaction (sample) is shown in the column. The association rules mining required the binary form of DiE_{norm} and DiM_{norm}. The row-wise (i.e., gene-wise) mean values of DiE_{norm} were normalized using the zero-mean normalization procedure, and they became zero. When discretizing expression data, any value greater than 0 denoting up-regulation was converted to 1 (denoted as UpR), while any value less than 0 denoting down-regulation (denoted as DwR) was converted to 0. The methylation data, on the other hand, turned any value greater than 0 indicating hyper-methylation (denoted as $HperM$) into 0 and any value less than 0 implying hypo-methylation (denoted as $HpoM$) into 1.

The following equations, eqn (3.5) and (3.6) respectively, define the aforementioned discretization method for the expression and methylation datasets:

$$DDiE_{norm}(i,j) = \begin{cases} 1, & \text{if } DiE_{norm}(i,j) > 0, \\ 0, & \text{if } DiE_{norm}(i,j) < 0 \end{cases} \tag{3.5}$$

$$DDiM_{norm}(i,j) = \begin{cases} 1, & \text{if } DiM_{norm}(i,j) < 0, \\ 0, & \text{if } DiM_{norm}(i,j) > 0, \end{cases} \tag{3.6}$$

where the discretized expression and methylation data matrices are denoted by $DDiE_{norm}$ and $DDiM_{norm}$, respectively. For i and j, the range of values is 1-n and 1-m, respectively. Then, all of the discretized matrices that were generated were transposed as follows:

$$DDiTE_{norm} = t(DDiE_{norm}), \tag{3.7}$$

and

$$DDiTM_{norm} = t(DDiM_{norm}). \tag{3.8}$$

The transposed discretized expression data (designated by $DDiTE_{norm}$ in eqn (3.7) and methylation data (designated by $DDiTM_{norm}$ in eqn (3.8) were assembled into a single binary matrix during the post-discretization (abbreviated by $PDiD_{em}$) with the size of $[m \times (2 * n)]$. Four categories of genes, namely (i) up-regulated and hypo-methylated genes, (ii) up-regulated and hyper-methylated genes, (iii) down-regulated and hyper-methylated genes, and (iv) down-regulated and hypo-methylated genes, were identified as a result of the integration of the expression and methylation data. The first and third categories of genesets (i.e., categories represented by (i) and (iii)) were chosen because gene expression and methylation are inversely correlated. As previously stated, the size of $PDiD_{em}$ is $[m \times (2 * n)]$. The column length (gene area) of the post-discretization is twice that of the transposed discretized expression/methylation matrix. The first half of the $PDiD_{em}$ column vector represents type (i) up-regulation and hypo-methylation, while the second half represents type (iii) down-regulation and hyper-methylation.

As a result, if a specific cell/house value (for example, cell at the jth sample and the ith gene) of the transposed discretized expression data matrix $DDiTE_{norm}$ is 1 (i.e., called up-regulated), and the same cell/house value of the transposed discretized methylation data matrix $DDiTM_{norm}$ is 1 (i.e., termed as hypo-methylation), then we put the symbol "1" in the same cell or location as the first half of the post-discretized matrix because it meets the criteria for category (i) up-regulation and hypo-methylation (i.e., the cell at the jth sample and the ith gene of $PDiD_{em}$ is considered as the first joint condition of eqn (3.9) that denoted type (i) both up-regulated and hypo-methylated genes. At the same time, we also placed the symbol "0" in the same cell/location of the second half of the post-discretized matrix (i.e., cell at the jth sample and the ith gene of $PDiD_em$ seen as the first joint condition of eqn (3.10), which is just the negation of "1".

In contrast, when both transposed discretized scores for the same cell or house were 0 (down-regulation and hyper-methylation), the resulting post-discretized value for the second half of the post-discretized matrix would be 1 (see the second joint condition of eqn (3.10), while the corresponding value for the first half of the post-discretized matrix would automatically be the inverse of that (i.e., 0) (see the second joint condition of eqn (3.9). Additionally, the post-discretized values for the first and second halves would be 0 for any possible combinations of the transposed discretized expression value and the transposed discretized methylation value, such as (0 and 1) and (1 and 0).

$$PDiD_{em}(j,i) = \begin{cases} 1, & \text{if } DDiTE_{norm}(j,i) == 1 \,\&\, DDiTM_{norm}(j,i) == 1, \text{ (UpR and HypoM)} \\ 0, & \text{if } DDiTE_{norm}(j,i) == 0 \,\&\, DDiTM_{norm}(j,i) == 0, \text{ (DwR and HyperM)} \\ 0, & \text{otherwise}, \end{cases} \tag{3.9}$$

and

$$PDiD_{em}(j,i+n) = \begin{cases} 0, & \text{if } DDiTE_{norm}(j,i) == 1 \,\&\, DDiTM_{norm}(j,i) == 1, \\ 1, & \text{if } DDiTE_{norm}(j,i) == 0 \,\&\, DDiTM_{norm}(j,i) == 0 \\ 0, & \text{otherwise}. \end{cases} \tag{3.10}$$

However, in Figure 3.1, we provided several examples of the aforementioned computations. After the post-discretization process, we transposed the resulting post-discretized matrix in preparation for the following steps.

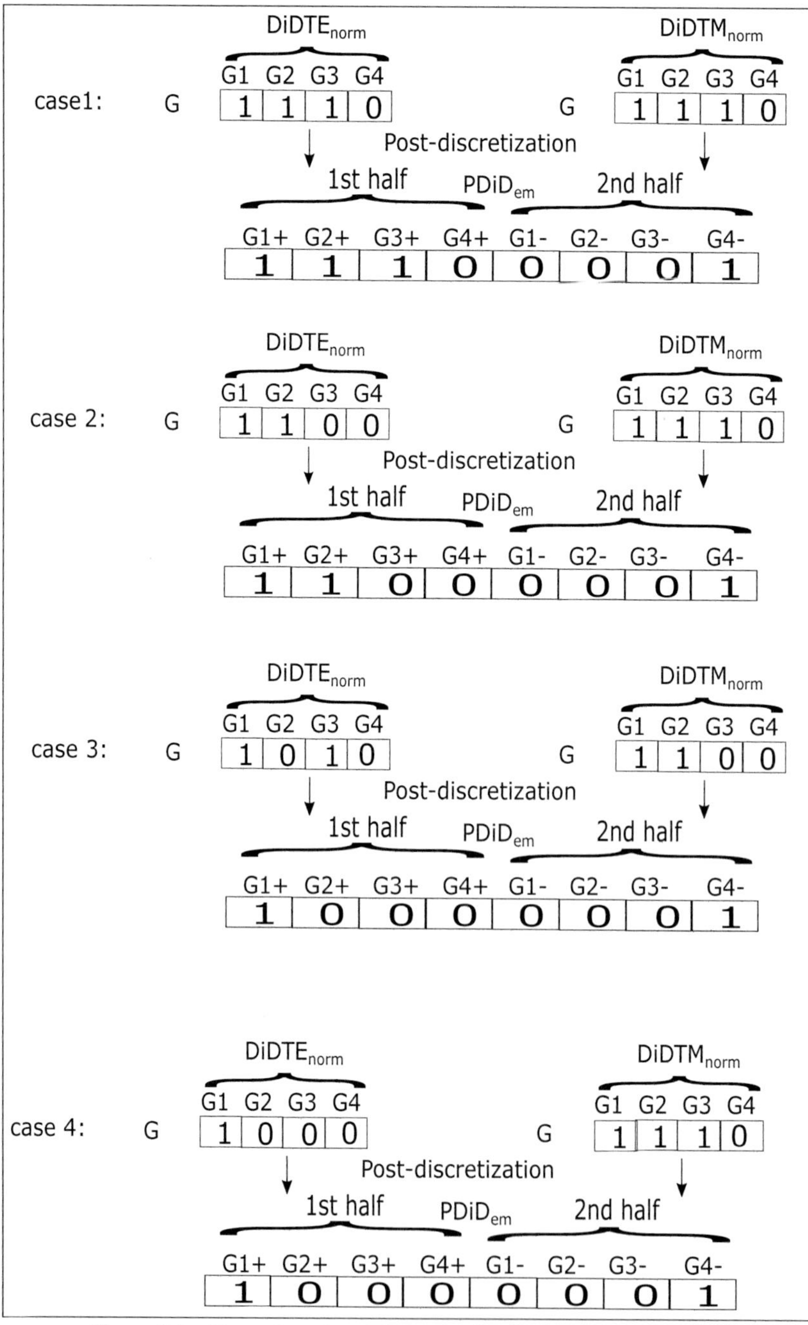

Figure 3.1 Examples of post-discretization in the proposed method.

3.3.3 Proposed association rule mining approach

The conventional concept of employing a static support threshold and a static confidence threshold — which are usually used to preserve these thresholds static across all itemsets, or genesets — was revised by the $MOOVARM$ approach. The weighted shortest distance dependent multiple minimum support threshold, multiple minimum confidence threshold, and multiple minimum lift threshold were used in our method after post-discretization in place of the static support threshold and static confidence threshold. The fusion of the gene expression, methylation, and protein–protein interaction profiles produced those various minimal thresholds. In place of separate datasets like gene expression or DNA methylation or any other data, the $MOOVARM$ technique worked on three distinct types of profiles simultaneously: methylation, protein–protein interaction, and gene expression. This resulted in multi-objective multi-prolific association rules. This $MOOVARM$ approach consisted of the following phases: The frequency of each gene (item) in the post-discretized data was calculated, along with the WV_{msc}, WV_{mcc}, and WV_{mlc} scores. Additionally, the geneset-tree ($GSTR$) was formed, genesets (itemsets) were generated, the D_bVS, D_bVC, and D_bVL were determined, and the top significant relation-dependent association rules were produced.

The binary matrix $PDiD_{em}$ was converted into the transactional matrix $TRDiM$, which contained transactions corresponding to multiple gene IDs per transaction, in the first phase. TR_n stood for the number of transactions that were present in $TRDiM$. It was necessary to describe both the user-mentioned minimal support cutoff ($UD_{min}S$) and the user-mentioned minimum confidence cutoff ($UD_{min}C$). The user-defined minimum lift cutoff, $UD_{min}L$, was left at a value of 1. Next, the frequency of each gene in the $TRDiM$ dataset was calculated. Genes were regarded as common if their frequency was more than or equal to UD_{minS}. The frequently expressed genes were organized from most frequent to least frequent (high to low order). The produced cutoff $WV_{msc}(;)$ was calculated in the second stage for each pair of genes by fusing $H(;)$, $CECM_{exm}(;)$, and $UD_{min}S$. Similarly, $WV_{mcc}(;)$ was computed by integrating $H(;)$, $CECM_{exm}(;)$, and $UD_{min}C$; and $WV_{mlc}(;)$ was evaluated by applying $H(;)$, $CECM_{exm}(;)$, and $UD_{min}L$, respectively. The geneset-tree ($GSTR$) was initially obtained in the following two phases, and the significant genesets were then constructed sequentially by adhering to the methods applied in the conventional $FP - Growth$ association rule mining approach. The distance-based cutoff (i.e., D_bVS,

D_bVC, and D_bVL) scores were then evaluated for each resultant geneset by the initially computed WV_{msc}, WV_{mcc}, and WV_{mlc} matrices, respectively, in the fifth phase. The support of each generated geneset was first determined in the last step. The frequently detected genesets (i.e., those whose support scores were superior to or were equivalent to the corresponding individual D_bVS threshold rather than the user-specified support threshold $UD_{min}S$) were then identified. The rules were then obtained with regard to the frequent genesets, and the confidences and lifts associated with each rule were calculated. We limited our selection of rules from the aforementioned collection to those whose confidence and lift scores exceeded or were equivalent to their respective D_bVC and D_bVL cutoffs rather than $UD_{min}C$ and $UD_{min}L$.

Figure 3.2 shows the flowchart for the proposed $MOOVARM$ rule mining technique and all model parameter symbols are defined in Table 3.1.

3.4 Multi-criteria (Multi-objective Optimization) Decision-Making Technique

The process to choose the best alternative from a list of finite alternatives with respect to several criteria is known as multi-criteria decision-making ($MCDM$) [47]. The $MCDM$ technique helps the decision-maker choose the optimal option in a conflicting circumstance and has numerous applications in a variety of sectors, including economy, management, engineering, and medical diagnostics, among others.

The decision-maker takes into account how many factors were important in achieving the goal under the process of decision-making. Some of those requirements conflict with one another; some are maximized, while others are minimized. Different $MCDM$ techniques, including $MAXMIN$ [48], $MAXMAX$ [48], AHP [45], $ELECTRE$ [46], $TOPSIS$ [41], etc., are used to solve these kinds of problems. These techniques are seen as a decision-making process based on the problem behaviors, such as ranking, scoring, selecting, and ordering, and the surrounding environment, such as the type and size of the accessible data, processing/execution time, internal consistency, and logical relationships. The $TOPSIS$ approach is the best $MCDM$ method in two situations: first, when there are issues with the huge number of criteria and alternatives; and second, when there are large amounts of objective data available. The positive ideal alternative that had the extreme performance on each criterion was first identified through the $TOPSIS$ approach. It also discovered the negative ideal alternative that led to the worst

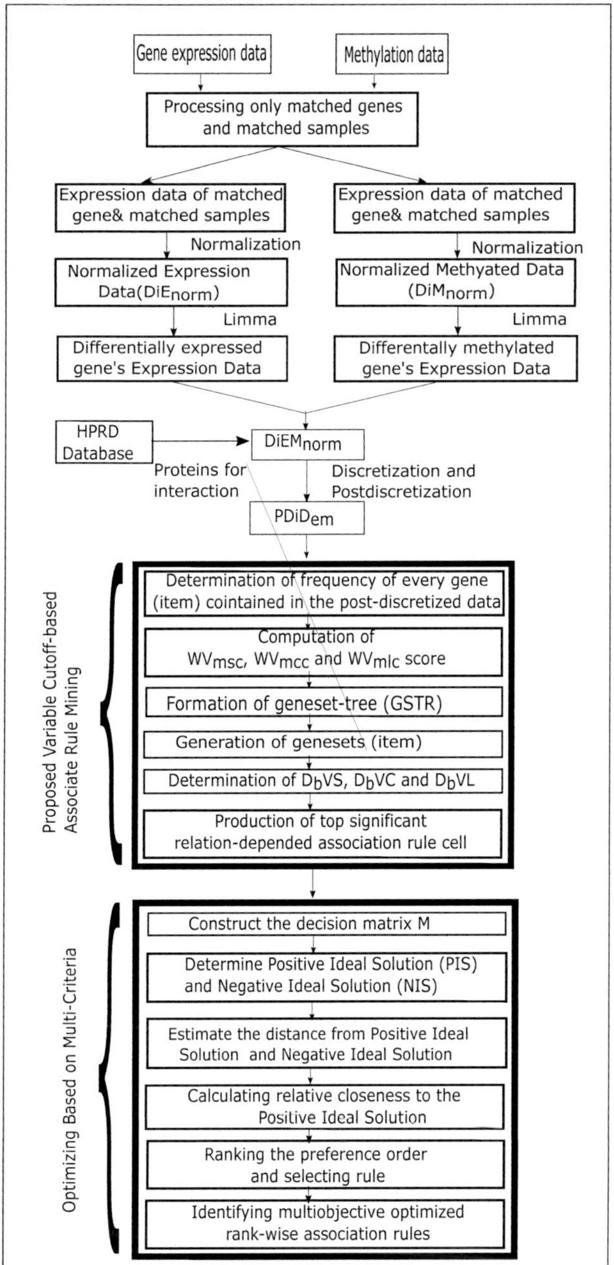

Figure 3.2 Flowchart of the proposed rule mining method.

Table 3.1 Model parameters.

Symbol	Definition
$CECM_{exm}$	Co-expression and co-methylation values
G	Gene vector
norm	Normalization technique
PCB	Pearson's correlation coefficient
G_{ex}	Vectors consisting of the expression and methylation values
D_{simt}	Dissimilarity scores
$HPRD$	Human Protein Resource Database
$DijStP(p, q)$	The weighted distance for every gene-pair (p, q) that contained the interactions in their corresponding protein levels among each other
W_eSD	Weighted shortest distance matrix
W_eSD_{mx}	Maximum weighted shortest distance
W_eSD_{mn}	Minimum weighted shortest distance
W_eSD_{avg}	Average weighted shortest distance
WV_{msc}	Minimum support threshold
WV_{mcc}	Minimum confidence threshold
WV_{mlc}	Minimum lift threshold
$UDminS$	User-defined minimum support threshold
$UD_{min}C$	User-mentioned minimum confidence cutoff
$UD_{min}L$	User-defined minimum lift cutoff
med	Median value
MAD	Median absolute deviation
GS	geneset
$GSTR$	Geneset-tree
ngp	The total number of possible gene pairs
D_bVS	Distance-based variable supports threshold
D_bVC	Distance-oriented variable confidence
D_bVL	Distance-oriented variable lift
$DiEM_{norm}$	Set of normalized genes
DiE_{norm}	Normalized expression
DiM_{norm}	Normalized methylation data-matrices
UpR	Up-regulated
DwR	Down-regulated
$HpoM$	Hypo-methylated gene
$DDiE_{norm}$	Discretized expression
$DDiM_{norm}$	Methylation data-matrices
$DDiTE_{norm}$	Transposed discretized expression data
$DDiTM_{norm}$	Methylation data
$PDiD_{em}$	Binary matrix
$TRDiM$	Transactional matrix
TR_n	Number of transactions
M_{nm}	Decision matrix
p_{ij}	Choice value
PIS^+	Positive ideal solution
NIS^-	Negative ideal solution

Algorithm 1 $MOOVARM$

Input: Gene expression (EX), DNA methylation data (Mt), and protein–protein interaction (PPI) data.

Output: List of rank-wise multi-objective optimized association rules.

1: **procedure** MOOVARM(EX, Mt, PPI)
2: Find the matched genes and matched samples between EX and Mt, and choose only them for EX/Mt.
3: Normalize EX/Mt by zero-mean normalization.
4: Identify differentially expressed genes from EX and differentially methylated genes from Mt; and intersect them and finally choose those intersected genes that have interactions in $HPRD$ database (denoted as $DiEM_{norm}$ geneset).
5: Discretize the EX/Mt subdata having $DiEM_{norm}$ geneset into $DDiE_{norm}(:)$ and $DDiM_{norm}(:)$ respectively, and post-discretize them together into a single matrix, $PDiD_{em}$ (see eqn (3.5)–(3.10)).
6: Transpose $PDiD_{em}$ into the transactional matrix $TRDiM$.
7: Generate frequent genesets GS from $TRDiM$:
8: **for** each gene $g_i \in TRDiM$ **do**
9: **if** frequency(g_i) $\geq UD_{min}S$ **then**
10: $GS \leftarrow g_i$
11: **end if**
12: **end for**
13: Determine WV_{msc}, WV_{mcc}, and WV_{mlc} cutoff scores for each pair of gene.
14: Form geneset-tree ($GSTR$) and then generate important geneset by FP-Growth rule mining method.
15: Distance based cutoffs (i.e., D_bVS, D_bVC, and D_bVL) scores were evaluated for every resultant geneset using $WV_{msc}(:)$, $WV_{mcc}(:)$, and $WV_{mlc}(:)$ scores, successively. Produce top significant relation-dependent association rules (see eqn (3.2)–(3.4)).
16: Develop the decision matrix M according to confidence, support, lift, and average WeSD value of rules.
17: Determine the positive ideal solution (PIS^+) and negative ideal solution (NIS^-) (see eqn (3.11) and (3.12)).
18: Calculate the distance (DIS_i^+) using PIS^+ and the distance (DIS_i^-) from NIS^- of each alternatives (see eqn (3.13) and (3.14)).
19: Compute the relative closeness (S_i) to the positive ideal solution of each alternative (see eqn (3.15)).
20: Rank the preference order according to relative closeness and select the alternative that is close to 1. Thereafter, rank the alternative depending upon S_i score in descending order.
21: **end procedure**

results for each criterion. The solution that maximizes the benefit criterion and minimizes the cost criterion was the positive ideal solution, whereas the solution that maximizes the cost criterion and minimizes the benefit criterion was the negative ideal solution. The approach then determines the alternative based on how close to the ideal solution is found on the positive aspect and how far away is found on the negative aspect. The information of the criteria used in the traditional $TOPSIS$ technique was gathered from quantitative data and expert opinions, whereas the created solution focused on evaluation, prioritization, and selection.

Through the subsequent steps, it will determine ranking and relative closeness:

Step 1: Building the decision matrix.
Assume that $M = (p_{ij})_{nxm}$ is a decision matrix where p_{ij} indicates the choice value of the ith alternative and the jth criteria.
Step 2: Figure out the positive ideal solution (PIS^+) and negative ideal solution (NIS^-).
The following is referred to as the positive ideal solution (PIS^+):

$$PIS^+ = \{p_1^+, p_2^+, ..., p_m^+\} = \\ \{(max_i(p_{ij})\,|j \in K), (min_i(p_{ij})\,|j \in L)\}. \tag{3.11}$$

The negative ideal solution (NIS^-) is represented in the following way:

$$NIS^- = \{p_1^-, p_2^-, ..., p_m^-\} = \\ \{(min_i(p_{ij})\,|j \in K), (max_i(p_{ij})\,|j \in L)\}, \tag{3.12}$$

where K relates to the benefit criteria, whereas L relates to the cost criteria.
Step 3: Determine the distance from positive ideal sSolution and negative ideal solution.
In the following way, the distance of the ith alternative from the positive ideal solution DIS_i^+ was determined:

$$DIS_i^+ = \left(\sum_{j=1}^{m}\left(p_j^+ - p_{ij}\right)\right), i = 1, 2, ..., n, \tag{3.13}$$

whereas the following equation was used to calculate the distance of the ith choice from the negative ideal solution DIS_i^-:

$$DIS_i^- = \left(\sum_{j=1}^{m}\left(p_{ij} - p_j^-\right)\right), i = 1, 2, ..., n. \tag{3.14}$$

Step 4: Compute the relative closeness to the positive ideal solution S_i through the following:

$$S_i = \frac{DIS_i^+}{DIS_i^+ + DIS_i^-}, \text{ where } 0 < S_i < 1, i = 1, 2, ..., n. \tag{3.15}$$

Step 5: Rank the alternatives and choose the closest alternative to 1. Rank the alternatives in decreasing order based on their S_i scores.

Notably, Algorithm 1 contains the main components of the proposed algorithm MOOVARM.

3.5 Experimental Dataset and Results

An integrative dataset from high-grade soft tissue sarcomas (*HSTS*) profiles for DNA methylation and gene expression (NCBI ID: GSE52392) [33], [34] was used in the experiment. The initial methylation profile comprises 27,578 methylation probes, whereas the initial gene expression profile has 48,645 genes. Notably, we chose samples that included both of these values. This was made up of two types of samples: (i) normal tumor cell line ($nrTCL$) (i.e., control samples) and (ii) undifferentiated pleomorphic liposarcoma ($UdPLs$) (diseased samples). There are 13 $nrTCL$ samples and 13 $UdPLs$ samples in the profile. Then, we choose the 12,438 genes that were matched in terms of both methylation and expression levels.

Initially, for this experiment, we chose the genes based on their DNA methylation and expression values. Since genes that had more than one single probe for methylation and expression profiles, we first filtered out the probes that included missing values and after that for both the expression and methylation data, we calculated the median absolute deviation (MAD) of the raw intensity values across all samples for each probe. In the case of the expression data, only the probe associated with a gene that has the highest MAD value among all probes linked to that gene is selected for analysis. The other probes related to that gene are excluded from the analysis of the expression dataset. This procedure is applied consistently to all genes within the expression dataset. A similar approach is adopted for the methylation data, where redundant methylation probes are discarded, ensuring that only the most relevant probe for each gene is analyzed in the methylation dataset. The Limma R tool [38] was then used on each probe to determine whether or not there was differential methylation or expression. All remaining probes for each gene were simply deleted from the methylation profile and the expression profile, keeping only the probe with the highest significance (minimum

Table 3.2 The ranks of the evolved rules prior to the optimization and after optimization in MOOVARM along with several rule interestingness measures, confidence, support, and lift as well as average WeSD and relative score.

$Rank_{prevMOO}$	Rule	Confidence	Support	Lift	Avg WeSD	Relative score	$Rank_{afterMOO}$
1	STAT3+→ MAPK3+	0.009	0.003461538	0.0195	0.00339	0.3596	2
2	TP53-→ MAPK3+	0.008888889	0.003076923	0.01925926	0.00408	0.3547	6
3	MAPK3+,TP53-→ STAT3+	0.00875	0.002692308	0.02275	0.00543	0.346	15
4	STAT3+,TP53-→ MAPK3+	0.01	0.002692308	0.02166667	0.003735	0.36	1
5	ESR1+→ MAPK3+	0.008333333	0.001923077	0.01805556	0.00357	0.3565	4
6	JUN+,STAT3+ → MAPK3+	0.008571429	0.002307692	0.01857143	0.00388	0.355	5
7	JUN+,FYN+→ MAPK3+	0.01	0.002307692	0.02166667	0.008	0.3354	22
8	STAT3+,FYN+→ MAPK3+	0.01	0.002692308	0.02166667	0.00751	0.3384	19
9	JUN+,TP53-→ MAPK3+	0.008571429	0.002307692	0.01857143	0.004225	0.3525	7
10	JUN+,STAT3+,TP53 → MAPK3+	0.01	0.001923077	0.02166667	0.003946667	0.3588	3
11	FYN+,TP53→ MAPK3+	0.01	0.002307692	0.02166667	0.007855	0.3361	21
12	JUN+,AR+→ MAPK3+	0.01	0.001923077	0.02166667	0.00567	0.3474	14
13	JUN+,AR+→ TP53-	0.01	0.001923077	0.02888889	0.0072	0.3399	18
14	TP53-,AR+→ JUN+	0.01	0.001923077	0.026	0.00541	0.3494	13
15	JUN+,AR+→ MAPK3+,TP53-	0.01	0.001923077	0.0325	0.006435	0.3444	17
16	MAPK3+,JUN+,AR+ → TP53-	0.01	0.001923077	0.02888889	0.00616	0.3454	16
17	TP53-,AR+→ MAPK3+,JUN+	0.01	0.001923077	0.0325	0.0054675	0.3496	12
18	MAPK3+,TP53-,AR+→ JUN+	0.01	0.001923077	0.026	0.005063333	0.3514	9
19	JUN+,TP53-,AR+→ MAPK3+	0.01	0.001923077	0.02166667	0.00514	0.351	11
20	GRB2→ STAT3-	0.01	0.002692308	0.02888889	0.01145	0.324	23
21	FYN→ TP53+	0.01	0.003076923	0.026	0.01186	0.3228	24
22	ESR1-,FYN→ TP53+	0.01	0.002307692	0.026	0.00783	0.337	20
23	ESR1-,MAPK3→ STAT3-	0.01	0.001923077	0.02888889	0.005065	0.3516	8
24	STAT3-,MAPK3→ ESR1-	0.01	0.001923077	0.02888889	0.005155	0.3511	10

* "+" denotes up-regulated and hypo-methylated genes, and "-" represents down-regulated and hyper-methylated genes.

corrected p-value) for the subsequent analysis. Next, we performed an intersection analysis using the Human Protein Resource Database ($HPRD$) [37] between the set of genes with differential methylation, the set of genes with differential expression, and the set of genes whose individual proteins interacted with one another. Here, we detected lots of such common genes. We created a protein–protein interaction (PPI) network for each dataset in which each protein represented a gene from the corresponding intersected set of genes. The genes were then rearranged according to the high to low order of the degree values for each node (gene) in the PPI network.

The discretization and post-discretization stages were then carried out, respectively. Then, we used the multi-objective optimized variable support based association rule mining that we had previously developed, $MOOVARM$. The resulting rules are shown in Table 3.2. Notably, we optimized the rules utilizing the four measures of each rule – confidence, support, lift, and $WeSD$ – by computing relative score in optimization, where confidence, support, and lift were used to maximize their values and $WeSD$ was used to minimize its value. Then, we sorted the rules from high to low based on the relative score. According to the top rule for the $HSTS$ dataset, {STAT3+, TP53- $\rightarrow$ MAPK3+ }, if gene $STAT3$ is both up-regulated and hypo-methylated and gene $TP53$ is both down-regulated and hyper-methylated, it is likely that gene $MAPK3$ will be both up-regulated and hypo-methylated. The confidence, support, lift, avg $WeSD$, and relative score values for this rule are 0.01, 0.00269, 0.02275, 0.00543, and 0.36, respectively. Prior to optimization, it had a ranking of 4, but now it is in the first rank since it has the best relative score among all the rules. The next four highest ranked optimized rules are {STAT3+$\rightarrow$ MAPK3+}, {JUN+,STAT3+,TP53-$\rightarrow$MAPK3+ }, {ESR1+$\rightarrow$ MAPK3+}, and {JUN+,STAT3+$\rightarrow$ MAPK3+}, with relative scores of 0.3596, 0.3588, 0.3565, and 0.355, respectively (in Table 3.2). The ranks before and after optimization for these top genes produced by MOOVARM are detailed in Table 3.2 along with all information about the various rule interestingness measures, $WeSD$, relative scores.

We employed and ran the PAM classifier to compare the classification performance of the various rules derived by MOOVARM, Apriori, and Eclat rule mining algorithms towards the samples in order to validate the significance of each of the top 10 rules (in Table 3.2) generated from $DTFP\text{-}Growth$. For this reason, we only took into account the associated features (genes) with both sides of each individual top 10 rule. We then performed a 10-fold cross-validation using the PAM classifier and the default parameters on the data to assess the significance of the combination of all the genes leading to the rule. On each occasion, the entire process was repeated 10 times. The classification accuracy results of top 10 rules of the $HSTS$ dataset are shown in Table 3.3. Similar to this, Tables 3.4 and 3.5 displayed the accuracy values of the top 10 rules according to the $Apriori$ and $Eclat$ algorithms, respectively. Figure 3.3 presents the graphical plot for the top 10 rules classification accuracy

Table 3.3 Top 10 rules of MOOVARM with their classification accuracy, specificity, sensitivity, and AUC values.

Rule ID	Rule	Avg classification accuracy (sd)	Avg specificity (sd)	Avg sensitivity (sd)	AUC	Std.overall.err.rate
1	{STAT3+,TP53- →MAPK3+}	85.38% (±0.0405)	85.38% (±0.0653)	83.84% (±0.0405)	0.926	0.04054202
2	{STAT3+→MAPK3+}	84.61% (±0.0181)	76.92% (±0.0243)	92.30% (±0.0243)	0.918	0.01813094
3	{JUN+,STAT3+, TP53- →MAPK3+}	89.23% (±0.0324)	89.23% (±0.0324)	89.23% (±0.0397)	0.967	0.03243362
4	{ESR1+→ MAPK3+}	80.76% (±0.0243)	76.92% (±0.0371)	76.15% (±0.0243)	0.83	0.02432521
5	{JUN+,STAT3+→ MAPK3+}	92.3% (±0.0162)	90% (±0.0228)	87.69% (±0.0324)	0.956	0.02432521
6	{TP53-→MAPK3+}	73.84% (±0.0397)	85.38% (±0.0606)	83.84% (±0.0648)	0.771	0.03972291
7	{JUN+, TP53- →MAPK3+}	83.46% (±0.0506)	95.38% (±0.0519)	71.53% (±0.0653)	0.877	0.05063697
8	{ESR1-, MAPK3+→STAT3+}	84.61% (±0.0268)	83.84% (±0.0371)	85.38% (±0.0371)	0942	0.02689253
9	{MAPK3+, TP53-, AR+→JUN+}	91.92% (±0.0326)	82.30% (±0.0537)	82.30% (±0.0537)	0.956	0.03268602
10	{STAT3+, MAPK3+→ESR1- }	84.61% (±0.0268)	83.84% (±0.0371)	85.38% (±0.0371)	0.942	0.02689253
	Average	0.8507	0.8477	0.8538	0.909	0.03085

* *sd denotes standard deviation; "+" denotes up-regulated and hypo-methylated genes; "-" represents down-regulated and hyper-methylated genes.

Table 3.4 Top 10 rules of *Apriori* with their classification accuracy, specificity, sensitivity, and AUC values.

Rule ID	Rule	Avg classification accuracy (sd)	Avg specificity (sd)	Avg sensitivity (sd)	AUC	Std.overall.err.rate
1	{STAT3-→GRB2-}	84.23% (±0.0121)	83.84% (± 0)	87.69% (±0.0243)	0.878	0.01216261
2	{STAT3-,MAPK3- →GRB2-}	83.46% (±0.0198)	76.92% (±0)	100% (±0.0397)	0.888	0.01986145
3	{MAPK3-,GRB2- →STAT3-}	88.46% (±0.0198)	76.92% (±0)	100% (±0.0397)	0.888	0.01216261
4	{GRB2-→STAT3-}	81.15% (±0.0181)	85.38% (±0.0362)	76.92% (±0)	0.90	0.01813094
5	{ESR1-→MAPK3-}	76.53% (±0.0256)	76.92% (±0.0362)	76.15% (±0.0362)	0.815	0.02564103
6	{STAT3-→MAPK3-}	84.23% (±0.0218)	79.23% (±0.0436)	88.46% (±0)	0.915	0.02183255
7	{STAT3-,GRB2- →MAPK3-}	84.61% (±0.0218)	82.3% (±0.0362)	86.92% (±0.0243)	0.902	0.021832557
8	{GRB2-→MAPK3-}	78.84% (±0.0268)	74.61% (±0.0324)	83.07% (±0.0324)	0.817	0.02689253
9	{ESR1-→GRB2-}	83.07% (±0.0371)	77.69% (±0.0324)	88.46% (±0.0606)	0.826	0.03715738
10	{JUN-→GRB2-}	75% (±0.0373)	65.38% (±0.0243)	87.69% (±0.0648)	0.777	0.0373779
	Average	0.8196	0.7792	0.8754	0.861	0.0233

* sd denotes standard deviation; "+" denotes up-regulated and hypo-methylated genes; "-" represents down-regulated and hyper-methylated genes.

Table 3.5 Top 10 rules of *Eclat* with their classification accuracy, specificity, sensitivity, and AUC values.

Rule ID	Rule	Avg classification accuracy (sd)	Avg specificity (sd)	Avg sensitivity (sd)	AUC	Std.overall.err.rate
1	{STAT3-,MAPK3- →GRB2- }	84.61% (±0.0218)	85.38% (±0.0362)	83.84% (±0.0243)	0.902	0.02183255
2	{STAT3-→GRB2-}	80.76% (±0.0243)	84.61% (±0.0362)	76.92% (±0.0324)	0.897	0.02432521
3	{MAPK3-,GRB2- →STAT3-}	84.61% (±0.0162)	85.38% (±0.0243)	83.84% (±0.0243)	0.903	0.01621681
4	{GRB2-→STAT3-}	81.15% (±0.0181)	85.38% (±0)	76.92% (±0)	0.90	0.01813094
5	{ESR1-→MAPK3-}	76.92% (±0.0256)	79.23% (±0.0362)	77.69% (±0.0362)	0.816	0.02564103
6	{STAT3-→GRB2-}	80.77% (±0.0243)	84.62% (±0.0362)	76.92% (±0.0324)	0.897	0.02432521
7	{STAT3-→MAPK3-}	84.23% (±0.0218)	77.69% (±0.0362)	90.77% (±0)	0.915	0.02183255
8	{GRB2-→MAPK3-}	84.62% (±0.0268)	76.92% (±0.0436)	92.30% (±0.0324)	0.817	0.02689253
9	{ESR1-,GRB2-}	83.46% (±0.0371)	76.15% (±0.0324)	90.77% (±0.0324)	0.826	0.03715738
10	{FYN+→TP53-}	75.38% (±0.0373)	68.46% (±0.0567)	82.30% (±0.0567)	0.714	0.0373779
	Average	0.8165	0.8038	0.8323	0.859	0.0254

* "+" denotes up-regulated and hypo-methylated genes; "-" represents down-regulated and hyper-methylated genes.

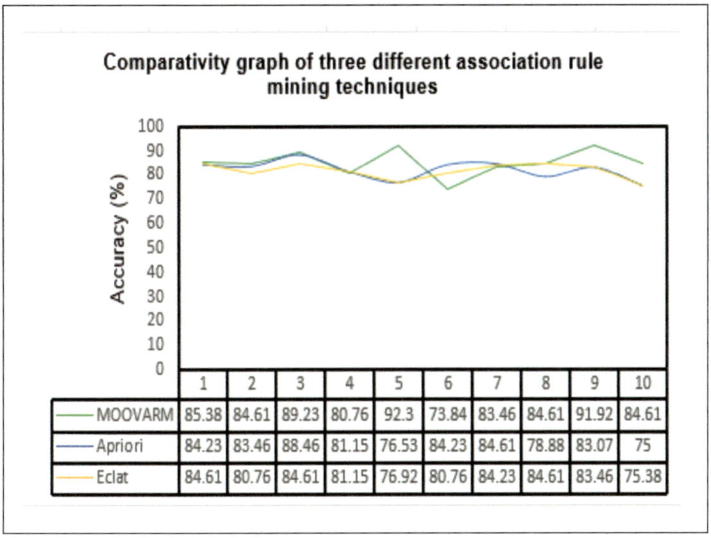

	1	2	3	4	5	6	7	8	9	10
MOOVARM	85.38	84.61	89.23	80.76	92.3	73.84	83.46	84.61	91.92	84.61
Apriori	84.23	83.46	88.46	81.15	76.53	84.23	84.61	78.88	83.07	75
Eclat	84.61	80.76	84.61	81.15	76.92	80.76	84.23	84.61	83.46	75.38

Figure 3.3 The comparative study between our proposed method $MOOVARM$ and other well-known related rule mining methods, $Apriori$ and $Eclat$, in terms of classification accuracy of generated top 10 rules obtained by (a) $MOOVARM$, (b) $Apriori$, and (c) $Eclat$.

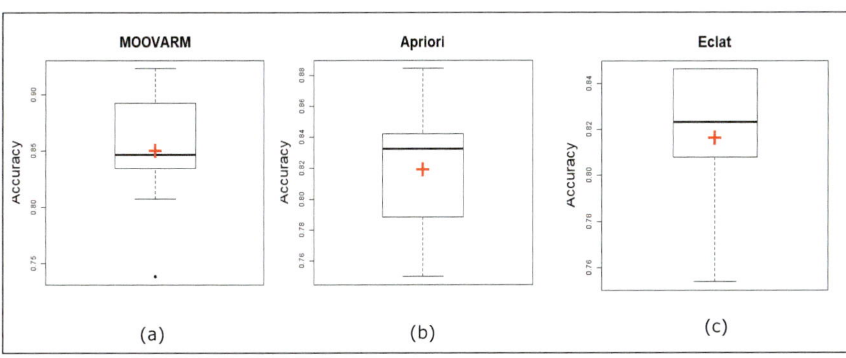

Figure 3.4 The comparative study between our proposed method $MOOVARM$ and other well-known related rule mining methods, $Apriori$ and $Eclat$, in terms of mean classification accuracy of generated rules obtained by (a) $MOOVARM$, (b) $Apriori$, and (c) $Eclat$ where "+" (red symbol) denotes the average classification accuracies and bold line signifies the median of them.

measurements derived using the three algorithms $MOOVARM$, $Apriori$, and $Eclat$. The proposed approach $MOOVARM$ clearly has a greater overall accuracy than the other two techniques based on the top 10 classification accuracy matrices. We also calculated the average classification accuracy of the top 10 rules, and the results were 85.08% ($\pm$0.03), 81.96% ($\pm$0.02), and 81.65% ($\pm$0.02) for the methods, $MOOVARM$, $Apriori$, and $Eclat$, respectively (in Figure 3.4). Additionally, the AUC in terms of Sensitivity vs. Specificity for classifying the corresponding disease using the genesets associated with the top two $MOOVARM$ rules (from Table 3.3) from the appropriate dataset was determined to be 0.926 and 0.920, respectively (see Figure 3.5). In addition, in conclusion, we found that the average AUC values of the top 10 rules from the same dataset using $MOOVARM$, $Apriori$, and $Eclat$ were 0.909, 0.861, and 0.859, respectively. Here, we utilized the open-source R package "pROC" [69] to show the ROC and AUC curves seen in Figure 3.5.

Furthermore, we used the participating genes from the top MOOVARM-generated rules to do KEGG pathway and gene ontology (GO) studies, after which we found the p-value significant GO-terms. While Table 3.8 provides the enrichment results for the gene ontology: biological processing (GO:BP) terms containing the resultant rules of MOOVARM, Tables 3.6 and 3.7 provide the results for the GO terms: molecular function (GO:MF) and cellular component (GO:CC), respectively. The KEGG pathways with the resulting MOOVARM rules are described in Table 3.9. An example is hsa05161:Hepatitis B KEGG pathway (p-value = 6.20E-06) contained 5 genes and 8 rules out of the 24 evolved rules that were produced by MOOVARM and are mentioned in Table 3.2. These five genes are GRB2, JUN, MAPK3, TP53, and STAT3, while these eight rules are {STAT3+$\rightarrow$MAPK3+}, {TP53-$\rightarrow$MAPK3+}, {MAPK3+,TP53-$\rightarrow$STAT3+}, {STAT3+,TP53-$\rightarrow$MAPK3+}, {JUN+,STAT3+$\rightarrow$MAPK3+}, {JUN+,TP53-$\rightarrow$MAPK3+}, {JUN+,STAT3+,TP53-$\rightarrow$MAPK3+}, and {GRB2-$\rightarrow$STAT3-}. In a similar manner, hsa05200:Pathways in cancer (p-value = 1.11E-05) was composed of 6 genes (AR, GRB2, JUN, MAPK3, TP53, and STAT3) and 15 developed rules. In case of the example of GO:BP terms, GO:0045893 positive regulation of transcription, DNA-templated (p-value = 5.31E-07) was associated with 6 genes (AR, JUN, MAPK3, TP53, ESR1, and STAT3) and 18 evolved rules, while for the example of GO:CC term, GO:0005654 nucleoplasm (p-value = 7.70E-05) was involved with 7 genes (AR, GRB2, JUN, MAPK3, TP53, ESR1, and STAT3) and 19 evolved rules. Five genes (AR, JUN, TP53, ESR1, and STAT3) and two derived rules

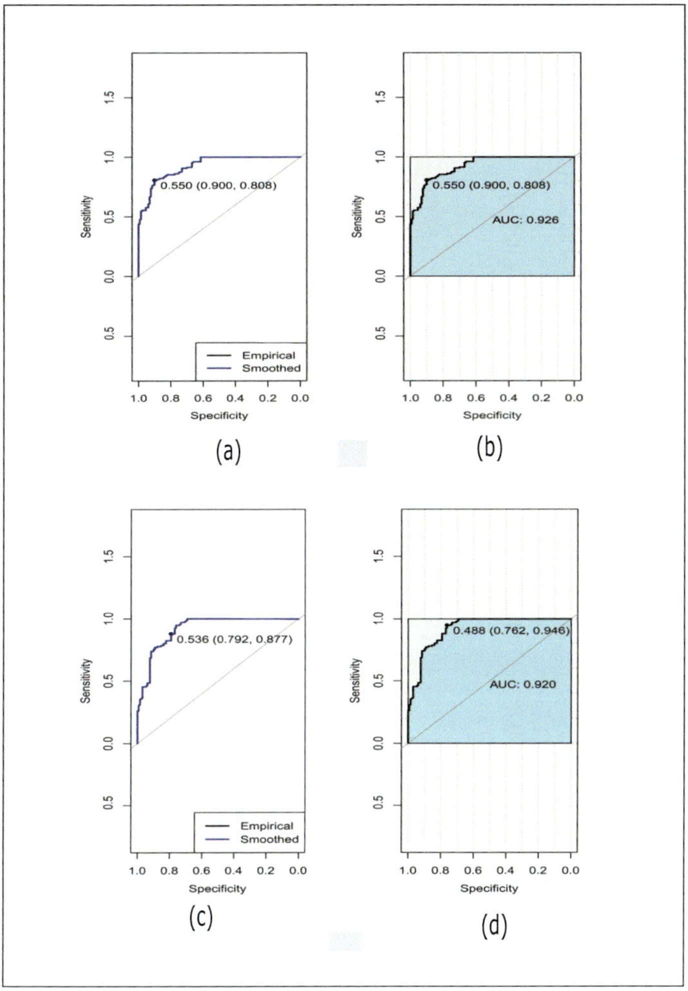

Figure 3.5 ROC curve of average accuracies and AUC in terms of sensitivity vs. specificity to classify the respective disease using the genesets belonging to the topmost two rules by MOOVARM (of Table 3.3) from the respective dataset. (a) ROC curve of average accuracies to classify the respective disease using the genesets belonging to the first rule, (b) AUC to classify the respective disease using the genesets belonging to the first rule, (c) ROC curve of average accuracies to classify the respective disease using the genesets belonging to the second rule, and (d) AUC to classify the respective disease using the genesets belonging to the second rule. Of note, the blue curve denotes the empirical line, while the light black line indicates smoothed control (X:Y=1:1) line. Here "0.550 (0.900, 0.808)" denotes the optimal threshold point (=0.550) closest to the top-left part of plot (for Rule 1) where the respective specificity and sensitivity are 0.900 and 0.808, respectively. Similarly, "0.536 (0.792, 0.877)" signifies the optimal threshold point (=0.536) closest to the top-left part of plot (for Rule 2) where the respective specificity and sensitivity are 0.792 and 0.877, respectively.

Table 3.6 Geneset enrichment result for gene ontology: molecular function (GO:MF) terms containing the resultant rules of MOOVARM.

GO:MF	*p*-value	Genes	Associated rules
GO:0008134 transcription factor binding	2.64E-06	AR, JUN, TP53, ESR1, STAT3	{JUN+,AR+→TP53-}, {TP53-,AR+→JUN+}
GO:0042802 identical protein binding	3.31E-06	FYN, GRB2, JUN, TP53, ESR1, STAT3	{GRB2-→STAT3-}, {FYN-→TP53+}, { ESR1-,FYN-→TP53+}
GO:0019899 enzyme binding	4.97E-06	AR, FYN, JUN, TP53, ESR1	{JUN+,AR+→TP53-},{TP53-,AR+→JUN+},{FYN-→TP53+},{ESR1-,FYN-→TP53+}
GO:0044212 transcription regulatory region DNA binding	6.68E-05	AR, JUN, TP53, STAT3	{JUN+,AR+→TP53-}, {TP53-,AR+→JUN+}
GO:0019903 protein phosphatase binding	2.84E-04	GRB2, TP53, STAT3	{STAT3+→MAPK3+}, {TP53-→MAPK3+}, {MAPK3+,TP53-→STAT3+}, etc*.
GO:0003700 transcription factor activity, sequence-specific DNA binding	3.18E-04	AR, JUN, TP53, ESR1, STAT3	{JUN+,AR+→TP53-}, {TP53-,AR+→JUN+}
GO:0003682 chromatin binding	4.03E-04	AR, JUN, TP53, ESR1	{JUN+,AR+→TP53-}, {TP53-,AR+→JUN+}
GO:0043565 sequence-specific DNA binding	9.17E-04	AR, JUN, TP53, ESR1	{JUN+,AR+→TP53-}, {TP53-,AR+→JUN+}
GO:0003677 DNA binding	0.002636593	AR, JUN, TP53, ESR1, STAT3	{JUN+,AR+→TP53-}, {TP53-,AR+→JUN+}
GO:0019901 protein kinase binding	0.00964876	GRB2, TP53, STAT3	{GRB2-→STAT3-}
GO:0005515 protein binding	0.010325405	AR, FYN, GRB2, JUN, MAPK3, TP53, ESR1, STAT3	{STAT3+→MAPK3+}, {TP53-→MAPK3+}, {MAPK3+,TP53-→STAT3+}, etc*.

Table 3.7 Geneset enrichment result for gene ontology: cellular component (GO:CC) terms containing the resultant rules of MOOVARM.

GO:CC	*p*-value	Genes	Associated rules
GO:0005654 nucleoplasm	7.70E-05	AR, GRB2, JUN, MAPK3, TP53, ESR1, STAT3	{STAT3+→MAPK3+}, {TP53-→MAPK3+}, {MAPK3+,TP53-→STAT3+}, etc*.
GO:0005634 nucleus	2.04E-04	AR, FYN, GRB2, JUN, MAPK3, TP53, ESR1, STAT3	{STAT3+→MAPK3+}, {TP53-→MAPK3+}, {MAPK3+,TP53-→STAT3+}, etc*.
GO:0005829 cytosol	2.13E-04	AR, FYN, GRB2, JUN, MAPK3, TP53, STAT3	{STAT3+→MAPK3+}, {TP53-→MAPK3+}, {MAPK3+,TP53-→STAT3+}, etc*.

Table 3.8 Geneset enrichment result for gene ontology: biological processing (GO:BP) terms containing the resultant rules of MOOVARM.

GO:BP	*p*-value	Gene	Associated rules
GO:0045893 positive regulation of transcription, DNA-templated	5.31E-07	AR, JUN, MAPK3, TP53, ESR1, STAT3	{STAT3+→MAPK3+}, {TP53-→MAPK3+}, {MAPK3+,TP53-→STAT3+}, etc*.
GO:0016032 viral process	3.31E-06	FYN, GRB2, TP53, MAPK3, STAT3	{STAT3+→MAPK3+}, {TP53-→MAPK3+}, {MAPK3+,TP53-→STAT3+}, etc*.
GO:0045944 positive regulation of transcription from RNA polymerase II promoter	1.28E-05	AR, JUN, MAPK3, TP53, ESR1, STAT3	{STAT3+→MAPK3+}, {TP53-→MAPK3+}, {MAPK3+,TP53-→STAT3+}, etc*.
GO:0008285 negative regulation of cell proliferation	4.25E-04	AR, JUN, TP53, STAT3	{JUN+,AR+→TP53-}, {TP53-,AR+→JUN+}
GO:0007586 aging	0.001951	GRB2, JUN, STAT3	{GRB2-→STAT3-}, {TP53-,AR+→JUN+}
GO:0042981 regulation of apoptotic process	0.003225	FYN, TP53, ESR1	{FYN-→TP53+}, {ESR1-,FYN-→TP53+}
GO:0006351 transcription, DNA-templated	0.004792	AR, MAPK3, TP53, ESR1, STAT3	{STAT3+→MAPK3+}, {TP53-→MAPK3+}, {MAPK3+,TP53-→STAT3+}, etc*.
GO:0060397 JAK-STAT cascade involved in growth hormone signaling pathway	0.006237	MAPK3, STAT3	{STAT3+→MAPK3+}
GO:0030154 cell differentiation	0.014472	FYN, GRB2, TP53	{FYN-→TP53+}
GO:0016310 phosphorylation	0.040956	MAPK3, STAT3	{STAT3+→MAPK3+}
GO:0006461 protein complex assembly	0.047374	MAPK3, TP53	{TP53-→MAPK3+}

Table 3.9 Geneset enrichment result for KEGG pathways containing the resultant rules of MOOVARM.

KEGG pathway	*p*-value	Genes	Associated rules
hsa05161:Hepatitis B	6.20E-06	GRB2, JUN, MAPK3, TP53, STAT3	{STAT3+→MAPK3+}, {TP53-→MAPK3+}, {MAPK3+,TP53-→STAT3+}, etc*.
hsa05200:Pathways in cancer	1.11E-05	AR, GRB2, JUN, MAPK3, TP53, STAT3	{STAT3+→MAPK3+}, {TP53-→MAPK3+}, {MAPK3+,TP53-→STAT3+}, etc*.
hsa05205: Proteoglycans in cancer	2.23E-05	GRB2, MAPK3, TP53, ESR1, STAT3	{STAT3+→MAPK3+}, {TP53-→MAPK3+}, {MAPK3+,TP53-→STAT3+}, etc*.
hsa05203:Viral carcinogenesis	2.46E-05	GRB2, JUN, MAPK3, TP53, STAT3	{STAT3+→MAPK3+}, {TP53-→MAPK3+}, {MAPK3+,TP53-→STAT3+}, etc*.
hsa04917:Prolactin signaling pathway	3.53E-05	GRB2, MAPK3, ESR1, STAT3	{STAT3+→MAPK3+}, {ESR1+→MAPK3+}, {GRB2-→STAT3-}, {ESR1-,MAPK3-→STAT3-}, {STAT3-,MAPK3-→ESR1-}
hsa05215:Prostate cancer	6.73E-05	AR, GRB2, MAPK3, TP53	{TP53-→MAPK3+}
hsa04915:Estrogen signaling pathway	9.58E-05	GRB2, JUN, MAPK3, ESR1	{ESR1+→MAPK3+}
hsa04660:T cell receptor signaling pathway	1.08E-04	FYN, GRB2, JUN, MAPK3	{JUN+,FYN+→MAPK3+}
hsa04722: Neurotrophin signaling pathway	1.70E-04	GRB2, JUN, MAPK3, TP53	{TP53-→MAPK3+}, {JUN+,TP53-→MAPK3+}
hsa04380: Osteoclast differentiation	2.20E-04	FYN, GRB2, JUN, MAPK3	{JUN+,FYN+→MAPK3+}
hsa05160:Hepatitis C	2.31E-04	GRB2, MAPK3, TP53, STAT3	{STAT3+→MAPK3+}, {TP53-→MAPK3+}, {MAPK3+,TP53-→STAT3+},

Table 3.9 *Continued*

KEGG pathway	*p*-value	Genes	Associated rules
{STAT3+,TP53-→MAPK3+}, {GRB2-→STAT3-}			
hsa05213: Endometrial cancer	0.00113869	GRB2, MAPK3, TP53	{TP53-→MAPK3+}
hsa05223:Non-small cell lung cancer	0.00131991	GRB2, MAPK3, TP53	{TP53-→MAPK3+}
hsa05221:Acute myeloid leukemia	0.00131991	GRB2, MAPK3, STAT3	{STAT3+→MAPK3+}, {GRB2-→STAT3-}
hsa04010:MAPK signaling pathway	0.00155686	GRB2, JUN, MAPK3, TP53	{TP53-→MAPK3+}, {JUN+,TP53-→MAPK3+}
hsa05210: Colorectal cancer	0.00161604	JUN, MAPK3, TP53	{TP53-→MAPK3+}, {JUN+,TP53-→MAPK3+}
hsa05214:Glioma	0.00177498	GRB2, MAPK3, TP53	{TP53-→MAPK3+}
hsa05212:Pancreatic cancer	0.00177498	MAPK3, TP53, STAT3	{STAT3+→MAPK3+}, {TP53-→MAPK3+}, {MAPK3+,TP53-→STAT3+}, {STAT3+,TP53- →MAPK3+}
hsa05220:Chronic myeloid leukemia	0.0021738	GRB2, MAPK3, TP53	{TP53-→MAPK3+}
hsa04919:Thyroid hormone signaling pathway	0.00536748	MAPK3, TP53, ESR1	{TP53-→MAPK3+}, {ESR1+→MAPK3+}

Table 3.9 *Continued*

KEGG pathway	p-value	Genes	Associated rules
hsa04071:Sphingolipid signaling pathway	0.00593268	FYN, MAPK3, TP53	{TP53-→MAPK3+}
hsa05162:Measles	0.00724782	FYN, TP53, STAT3	{FYN-→TP53+}
hsa04068:FoxO signaling pathway	0.00735406	GRB2, MAPK3, STAT3	{STAT3+→MAPK3+}
hsa04550:Signaling pathways regulating pluripotency of stem cells	0.0080066	GRB2, MAPK3, STAT3	{STAT3+→MAPK3+}
hsa04062:Chemokine signaling pathway	0.01384442	GRB2, MAPK3, STAT3	{STAT3+→MAPK3+}
hsa05216:Thyroid cancer	0.02902286	MAPK3, TP53	{TP53-→MAPK3+}
hsa05206:MicroRNAs in cancer	0.03102958	GRB2, TP53, STAT3	{GRB2-→STAT3-}
hsa05219:Bladder cancer	0.04081937	MAPK3, TP53	{TP53-→MAPK3+}
hsa04151:PI3K-Akt signaling pathway	0.04418067	GRB2, MAPK3, TP53	{TP53-→MAPK3+}

({JUN+,AR+→TP53-} and {TP53-,AR+→JUN+}) were connected to the GO:MF term, GO:0008134 transcription factor binding (p-value = 2.64E-06).

The relationship between directional signatures and disease discovery is connected to association rule mining. The top association rule is STAT3+, TP53- → MAPK3+, where STAT3 and TP53 have contrasting functions in regulation of cellular pathways. The activation role of STAT3 up-regulates the survival pathway, whereas p53 promotes the apoptotic pathway, according to a literature review. STAT3 plays a role in the growth of cancer cells and is associated with tumor malignancy. A well-known tumor suppressor gene is TP53. Apoptosis, DNA repair, and cell cycle arrest are three mechanisms by which TP53 protects against DNA damage. In cancer, p53 mutation is frequently seen, especially in late stages of malignancy [53]. The rule states that it is likely that subsequent genes (MAPK3+) will be expressed/methylated in specified manner together if antecedent genes (STAT3+, TP53-) express/methylate in a stated manner together. According to a literature review, MARK3 controls the growth and bone metastases of human breast cancer cells [54].

Here we have conducted literature mining to demonstrate the effectiveness of our top 10 association rules. Our first association rule, {STAT3+,TP53-→MAPK3+}, states that if antecedent genes (STAT3+, TP53-) express/methylate in a specific manner together, it is likely that subsequent genes (MAPK3+) will do the same. We have these three genes STAT3, TP53, and MAPK3 combined as a core target gene or the highest degree hub gene connected to numerous diseases including gastric cancer, type 2 diabetes mellitus, etc., according to Yang et al. [64], [65], and Liu et al. (2022) [66]. Similarly, the three genes of the second association rule, JUN, TP53, and MAPK3, are also connected to gastric cancer, according to Zu et al. (2021) [67]. The three genes related with rule 8 – ESR1, MAPK3, and STAT3 – were jointly identified by Rani et al. [68] as the top hub protein target genes in PPI network analysis. Therefore, we can draw the conclusion that the genes linked to our association rules have collectively performed several roles in current literature studies.

However, from a conceptual point of view, the $MOOVARM$ approach modifies the conventional idea of using a static support threshold and a static confidence threshold, which were typically used to maintain these thresholds constant across all itemsets (i.e., genesets) in the traditional algorithms like Apriori and Eclat. The association rule mining algorithm in $MOOVARM$ uses weighted shortest distance dependent multiple minimum support thresholds, multiple minimum confidence thresholds, and multiple minimum lift thresholds in place of static support thresholds and static confidence thresholds after post-discretization. Those multiple/dynamic minimum thresholds were calculated using a weighted shortest distance based scheme combination with the integration of the gene expression, methylation, and protein–protein interaction profiles. Instead of using individual datasets for gene expression, DNA methylation, or any other form of data, the $MOOVARM$ technique uses three distinct types of profiles – gene expression, methylation, and protein–protein interaction profiles together – to construct multi-objective multi-prolific association rules. Here we also used a multi-objective optimization technique called Topsis, also known as the multi-criteria decision making technique. It is the process of choosing the best alternative out of a group of finite options according to lots of different criteria. Here, we ranked the association rules based on multiple criteria (including weighted support, weighted confidence, and weighted lift) and selected the top-ranked association rules using the multi-objective optimization technique. In simple terms, traditional rule mining methods like Apriori and Eclat use static user-defined threshold values and do not optimize the ranking of estimated rules, whereas $MOOVARM$ uses dynamic thresholds and produces optimized association rules by using multi-objective optimization on a variety of objectives/criteria/rule-interestingness values for each individual rule.

In addition, we have used the top-optimized association rule STAT3+, TP53- MAPK3+ estimated by $MOOVARM$ (Table 3.3), which is a directed gene signature, to further illustrate the relationship between association rule mining and directional gene signature and their impact on disease identification. According to the rule, it is likely that subsequent genes will express and methylate in the specified manner together (e.g., MAPK3 will be up-regulated and hypo-methylated) if the antecedent genes express and methylate in the specified manner simultaneously (e.g., STAT3 is up-regulated and hypo-methylated, and TP53 is down-regulated and hyper-methylated). Since there are three participating genes in the associated rule, the combined effects of the rule leads in the creation of a directional three-gene signature.

3.6 Conclusion and Future Works

In this chapter, we indicated a novel associated rule mining technique called $MOOVARM$ (multi-objective optimised variable cutoff based association rule mining) for the purpose of identifying the most appropriate and important rule for multi-omics profiles. Based on the concept of multi-objective optimization and weighted shortest distance, we implemented and integrated the gene expression, methylation, and protein–protein interaction data to generate the intriguing rules for multi-omics profiles. We determined the positive ideal solution (PIS) and the negative ideal solution (NIS) for each geneset to achieve this. The PIS increased profit while reducing loss. As an alternative, NIS increased the loss and decreased the profit. Then, for each geneset in the multi-omics datasets, we estimated the distances $d+$ and $d-$ from PIS and NIS, respectively. The relative closeness of the PIS for ranking the genesets was then determined using these two distances. In the proposed approach, global relative closeness scores rather than single gene scores were obtained. The extracted (multi-objective optimized) rules of correlated genes were finally ranked using MOOVARM, which may have an important influence on disease classification performance when compared to state-of-the-art algorithms in disease discovery as well as therapeutic value. The drawback of this work is that MOOVARM only works with multi-omics RNA-seq/microarray datasets that include protein–protein interaction dataset, DNA methylation dataset, and gene expression dataset for the same set of patients/samples. MOOVARM is unable to process a single omic dataset. Additionally, without the use of matrix imputation before pre-filtering processes, our method could not be effective for single-cell sequencing data.

In our future work, we will use the advanced adding approach to integrate new datasets. Additionally, we are eager to continue applying our suggested strategy for locating hub genes in studies of multi-molecular regulation (i.e., the regulation of long non-coding RNAs, transcription factors, microRNAs, and target genes) in order to determine the directional optimized gene signatures. The methodology will also be applied to single-cell RNA sequencing and single-cell ATAC sequencing data to find directed gene signatures for cancer detection.

Furthermore, we have examined various sophisticated association rule mining works based on fuzzy or rough set theory; however, the outcome rules are insufficient, indicating that the outcomes are not always beneficial when employing fuzzy/rough set based association rule mining [55], [56]. Comparatively, the outcome rules are optimized and sufficient because our proposed method $MOOVARM$ applied a multi-objective optimization strategy. Therefore, it is not always advantageous to merely provide fuzzy or rough sets. Therefore, we believe that combining fuzzy/rough set based rule mining with multi-objective optimization will be an effective first step in enhancing the performance of fuzzy/rough theory based methods. In order to produce better and more efficient association rule mining from multi-omic data, we will expand our proposed framework by using fuzzy/rough theory and multi-objective optimization together in future work.

References

[1] S. Bandyopadhyay and et al., *A Survey and Comparative Study of Statistical Tests for Identifying Differential Expression from Microarray Data*, IEEE/ACM TCBB, 2014, DOI: 10.1109/TCBB.2013.147.

[2] S.Bandyopadhyay, and S.Mallik, *Integrating Multiple Data Sources for Combinatorial Marker Discovery*, IEEE/ACM Transactions on Computational Biology and Bioinformatics, 2016, DOI: 10.1109/TCBB.2016.2636207.

[3] A. Navarro, P. Yin, D. Monsivais, et al., *Genome-Wide DNA Methylation Indicates Silencing of Tumor Suppressor Genes in Uterine Leiomyoma*, PLoS One, vol. 7, no. 3, art. e33284, 2012.

[4] S. Mallik, et al., *Integrated Analysis of Gene Expression and Genome-wide DNA Methylation for Tumor Prediction: An Association Rule Mining-based Approach*, Proc. IEEE Symposium on Computational Intelligence in Bioinformatics and Computational Biology (CIBCB), Singapore, pp. 120–127, April 16, 2013, DOI: 10.1109/CIBCB.2013.6595397.

[5] S. Bandyopadhyay and M. Bhattacharyya, *A Biologically Inspired Measure for Coexpression Analysis*, IEEE/ACM Transactions on Computational Biology and Bioinformatics, vol. 8, no. 4, pp. 929-942, 2011.

[6] R. Agrawal, T. Imielinski and A. Swami, *Mining Association Rules between Sets of Items in large Databases*, Proc. ACM SIGMOD, New York, NY, USA: ACM, pp. 207–216, 1993.

[7] M.J. Zaki, *Scalable Algorithms for Association Mining*, IEEE Transactions on Knowledge and Data Engineering, 12(3): 372–390, 2000.

[8] R. Alves, et al., *Gene association analysis: a survey of frequent pattern mining from gene expression data*, Brief. Bioinformatics, 11, pp. 210-224, 2010.

[9] C. Creighton and S. Hanash, *Mining gene expression databases for association rules*, Bioinformatics, vol. 19, pp. 79-86, 2003.

[10] E. Georgii and et al., *Analyzing microarray data using quantitative association rules*, Bioinformatics, vol. 21, pp. 123-129, 2005.

[11] T. McIntosh and S. Chawla, *High confidence rule mining for microarray analysis*, IEEE/ACM Trans. Comput. Biol. Bioinformatics, vol. 4, pp. 611-623, 2007.

[12] J. Han, J. Pei, Y. Yin and R. Mao, *Mining Frequent Patterns without Candidate Generation: A Frequent-Pattern Tree Approach*, Data Mining and Knowledge Discovery, 8, 53-87, 2004.

[13] R. Martinez and et al., *GenMiner: mining non-redundant association rules from integrated gene expression data and annotations*, Bioinformatics, vol. 24, pp. 2643-2644, 2008.

[14] S. Mallik, A. Mukhopadhyay and U. Maulik, *RANWAR: Rank-Based Weighted Association Rule Mining from Gene Expression and Methylation Data*, IEEE Transactions on Nanobioscience, 14:1, pp. 59–66, 2015.

[15] G.D. Ramkumar and et al., *Weighted association rules: model and algorithm*, In Proceedings of the Fifth ACM SIGKDD International Conference on Knowledge Discovery and Data Mining. ACM, 1998.

[16] U. Maulik, S. Mallik, A. Mukhopadhyay and S. Bandyopadhyay, *Analyzing Gene Expression and Methylation Data Profiles using StatBicRM: Statistical Biclustering-based Rule Mining*, PLoS One, 10(4): e0119448, 2015.

[17] C.H. Cai and et al., *Mining association rules with weighted items*, In Proceedings of the International Database Engineering and Applications Symposium, IEEE Computer Society, Cardiff, Wales, UK, pp. 68-77, 1998.

[18] W. Wang and et al., *Efficient mining of weighted association rules*, In Proceedings of the sixth ACM SIGKDD international conference on Knowledge discovery and data mining, ACM, Boston, MA, USA, pp. 270-274, 2000.

[19] Y.-C. Liu, C.-P. Cheng and V.S. Tseng, *Discovering relational-based association rules with multiple minimum supports on microarray datasets*, Bioinformatics, vol. 27, no. 22, pp. 3142–3148, 2011, doi:10.1093/bioinformatics/btr526.

[20] U. Yun and J.J. Leggett, *WFIM: weighted itemset mining with a weight range and a minimum weight*, In Proceedings of the SIAM International Data Mining Conference, Society of Industrial and Applied Mathematics, Newport Beach, California, USA, pp. 270-274, 2005.

[21] F.Tao, *Weighted Association Rule Mining using Weighted Support and Significance Framework*, Proc. ACM SIGKDD, Washington D.C., USA, pp. 661–666, 2003.

[22] V. Tseng and et al., *UP-Growth: an efficient algorithm for high utility itemsets Mining*, In Proceedings of the 16th ACM SIGKDD International Conference on Knowledge Discovery and Data Mining. ACM, Washington, DC, USA, pp. 253-262, 2010.

[23] S.C. Madeira and A.L. Oliveira, *Biclustering algorithms for biological data analysis: a survey*, IEEE Trans. Comput. Biol. Bioinformatics, 1, 24-45, 2004.

[24] Y. Cheng and G.M. Church, *Biclustering of expression data. In Proceedings of the 8th International Conference on Intelligent Systems for Molecular Biology*, AAAI Press, Menlo Park, California, USA, pp. 93-103, 2000.

[25] J. Pei and et al., *MaPle: a fast algorithm for maximal pattern-based clustering*, In Proceedings of the 3rd IEEE International Conference on Data Mining, IEEE Computer Society, Melbourne, Florida, USA, pp. 259-266, 2003.

[26] D. Jiang, et al., *Cluster analysis for gene expression data: a survey*, IEEE Trans. Knowl. Data Eng., 16, 1370-1386, 2004.

[27] A. Thalamuthu, et al., *Evaluation and comparison of gene clustering methods in microarray analysis*, Bioinformatics, 22, 2405-2412, 2006.

[28] A. Prelic, et al., *A systematic comparison and evaluation of biclustering methods for gene expression data*, Bioinformatics, 22, pp. 1122-1129, 2006.

[29] M. Bhasin and G.P.S. Raghava, *SVM based method for predicting HLADRB1*0401 binding peptides in an antigen sequence*, Bioinformatics, 20, pp. 421-423, 2004.

[30] A. Paziewska, M. Dabrowska, K. Goryca, A. Antoniewicz, et al., *DNA methylation status is more reliable than gene expression at detecting cancer in prostate biopsy*, Br J Cancer, 12:111(4), pp. 781–9, 2014.

[31] F. Martella, *Classification of microarray data with factor mixture models*, Bioinformatics, 22, pp. 202-208, 2009.

[32] K.H. Liu and C.G. Xu, *A genetic programming-based approach to the classification of multiclass microarray datasets*, Bioinformatics, 25, 331-337, 2009.

[33] M. Renner, T. Wolf, H. Meyer, et al., *Integrative DNA methylation and gene expression analysis in high-grade soft tissue sarcomas*, Genome Biol., vol. 14, no. 12, 2013, doi: 10.1186/gb-2013-14-12-r137.

[34] P. Chudasama, M. Renner, M. Straub, et al., *Targeting Fibroblast Growth Factor Receptor 1 for Treatment of Soft-Tissue Sarcoma*, Clin Cancer Res., vol. 23, no. 4, pp. 962–973, 2017, doi: 10.1158/1078-0432.CCR-16-0860.

[35] S. Mallik, and Z. Zhao, *Towards integrated oncogenic marker recognition through mutual information-based statistically significant feature extraction: an association rule mining based study on cancer expression and methylation profiles*, Quantitative Biology, 5(4), pp. 302–327, 2017.

[36] S. Mallik, and Z. Zhao, *TrapRM: Transcriptomic and proteomic rule mining using weighted shortest distance based multiple minimum supports for multi-omics dataset*, IEEE International Conference on Bioinformatics and Biomedicine (BIBM), 13-16 Nov. 2017, pp. 2187–2194.

[37] S. Peri, J.D. Navarro, R. Amanchy, et al., *Development of human protein reference database as an initial platform for approaching systems biology in humans*, Genome Res., vol. 13, no. 10, pp. 2363–71, 2003.

[38] G. Smyth, *Linear models and empirical bayes methods for assessing differential expression in microarray experiments*, Stat Appl Genet Mol Biol., vol. 3, Article id. 3, 2004.

[39] T. Bhadra, S. Mallik, and S. Bandyopadhyay, *Identification of Multi-View Gene Modules using Mutual Information Based Hypograph Mining*, IEEE Transactions on Systems, Man, and Cybernetics: Systems, 2017 (accepted), DOI: 10.1109/TSMC.2017.2726553

[40] S. Mallik, A. Mukhopadhyay and U. Maulik, *Integrated Statistical and Rule–Mining Techniques for DNA Methylation and Gene Expression Data Analysis*, Journal of Artificial Intelligence and Soft Computing Research, 3(2), 2013.

[41] C.L. Hwang, K. Yoon, *Multiple Attribute Decision Making*,Springer–Verlag, Berlin, 1981.

[42] T. Bhadra, S. Mallik, and Ayan Mukherji , *DTFP–Growth: Dynamic Threshold Based FP–Growth Rule Mining Algorithm Through Integrating Gene Expression, Methylation and Protein-Protein Interaction Profiles*, IEEE Transactions on Systems, Man, and Cybernetics: Systems, 2018 (accepted), DOI: 10.1109/TSMC.2017.2726553

[43] Su Y, Blake-Palmer KG, Sorrell S, Javid B, Bowers K, Zhou A, Chang SH, Qamar S, Karet FE, *Human H+ATPase a4 subunit mutations causing renal tubular acidosis reveal a role for interaction with phosphofructokinase-1*, 2008, Am J Physiol Renal Physiol 295(4):F950-8

[44] Z. Liu , M.J. Macias, M.J. Bottomley , G. Stier, J.P. Linge , M. Nilges , P. Bork, M. Sattler, *The three-dimensional structure of the HRDC domain and implications for the Werner and Bloom syndrome proteins. Structure*, 1999, Research Support 7(12):1557-66

[45] T. L. Saaty, *How to Make a Decision: The Analytic Hierarchy Process*, Interfaces, vol. 24, no. 6, pp. 19-43, Nov.-Dec. 1994

[46] B. Roy and D. Vanderpooten. *The European school of MCDA: Emergence, basic features and current works*, Journal of Multi-Criteria Decision Analysis, 5(1):22–37, 1996

[47] S. Das, M.B. Kar, S. Kar, *Group multi-criteria decision making using intuitionistic multi-fuzzy sets*, Journal of Uncertainty Analysis and Applications 10(1), 1–16, 2013

[48] Y.Y. Shen, L.F. Mo,*The max–min approach to a relativistic equation*,Computers and Mathematics with Applications, Volume 58, Issues 11–12, December 2009, Pages 2131-2133

[49] Theilhaber J, Chiron M, Dreymann J, Bergstrom D, Pollard J., *Construction and optimization of gene expression signatures for prediction of survival in two-arm clinical trials*, BMC Bioinformatics, 2020 Jul 25;21(1):333, doi: 10.1186/s12859-020-03655-7, PMID: 32711453; PMCID: PMC7382041.

[50] Murphy RG, Gilmore A, Senevirathne S, O'Reilly PG, LaBonte Wilson M, Jain S, McArt DG, *Particle swarm optimization artificial intelligence technique for gene signature discovery in transcriptomic cohorts*, Comput Struct Biotechnol J., 2022, Sep 26;20:5547-5563, doi: 10.1016/j.csbj.2022.09.033, PMID: 36249564; PMCID: PMC9556859.

[51] Mahendran Nivedhitha, Durai Raj Vincent P. M., Srinivasan Kathiravan, Chang Chuan-Yu, *Machine Learning Based Computational Gene Selection Models: A Survey, Performance Evaluation, Open Issues, and Future Research Directions*, Frontiers in Genetics, Volume 11, 2020, ISSN: 1664-8021, doi: 10.3389/fgene.2020.603808.

[52] Ganguly, S., Mukherjee, V., *Frequency Stabilization of Isolated Hybrid Power System by a Novel Quasi-Oppositional Whale Optimization Algorithm*, Iran J Sci Technol Trans Electr Eng 44, 1467–1486 (2020), https://doi.org/10.1007/s40998-020-00341-5.

[53] Pham TH, Park HM, Kim J, Hong JT, Yoon DY, *STAT3 and p53: Dual Target for Cancer Therapy*, Biomedicines. 2020 Dec 21;8(12):637. doi: 10.3390/biomedicines8120637. PMID: 33371351; PMCID: PMC7767392.

[54] Du, Y., Zhang, J., Meng, Y. et al.,*MicroRNA-143 targets MAPK3 to regulate the proliferation and bone metastasis of human breast cancer cells*, AMB Expr 10, 134 (2020). https://doi.org/10.1186/s13568-020-01072-w

[55] S. Sharmila, S.Vijayarani, *Comparative Analysis Of Fuzzy Association Rule Mining Algorithms*, INTERNATIONAL JOURNAL OF SCIENTIFIC & TECHNOLOGY RESEARCH, VOLUME 8, ISSUE 11, NOVEMBER 2019, ISSN 2277-8616.

[56] Sandeep Kumar Singh, Mr.Ganesh Wayal, Mr.Niresh sharma, *A Review: Data Mining with Fuzzy Association Rule Mining*, International Journal of Engineering Research & Technology (IJERT) Vol. 1 Issue 5, July - 2012 ISSN: 2278-0181

[57] Mudi, J., Shiva, C.K. & Mukherjee, V., *Multi-verse Optimization Algorithm for LFC of Power System with Imposed Nonlinearities Using Three-Degree-of-Freedom PID Controller*, Iran J Sci Technol Trans Electr Eng 43, 837–856 (2019), https://doi.org/10.1007/s40998-018-0166-1.

[58] Mudi, J., Shiva, C.K., Vedik, B. et al., *Frequency Stabilization of Solar Thermal-Photovoltaic Hybrid Renewable Power Generation Using Energy Storage Devices*, Iran J Sci Technol Trans Electr Eng 45, 597–617 (2021), https://doi.org/10.1007/s40998-020-00374-w.

[59] Mudi, J., Shiva, C.K. and Mukherjee, V. (2021), *An Optimal Control of Integrated Hybrid Power System with FACTS Devices Using Student Psychology-Based Optimization Algorithm*, Adv. Theory Simul., 4: 2100147, https://doi.org/10.1002/adts.202100147.

[60] Mudi, J., Shiva, C.K. and Mukherjee, V. (2022), *Quasi-Oppositional Whale Optimization Optimized Load Frequency Stabilization of Hybrid Power Systems Integrated with Electric Vehicle*, Adv. Theory Simul., 5: 2100510, https://doi.org/10.1002/adts.202100510.

[61] Somnath Ganguly, Chandan Kumar Shiva, V. Mukherjee, *Frequency stabilization of isolated and grid connected hybrid power system models*, Journal of Energy Storage, Volume 19, 2018, Pages 145-159, ISSN 2352-152X, https://doi.org/10.1016/j.est.2018.07.014.

[62] Somnath Ganguly, Tarkeshwar Mahto, V. Mukherjee, *Integrated frequency and power control of an isolated hybrid power system considering scaling factor based fuzzy classical controller, Swarm and Evolutionary Computation*, Volume 32, 2017, Pages 184-201, ISSN 2210-6502, https://doi.org/10.1016/j.swevo.2016.08.001.

[63] Ganguly, S., Mukherjee, V., *Frequency Stabilization of Isolated Hybrid Power System by a Novel Quasi-Oppositional Whale Optimization Algorithm*, Iran J Sci Technol Trans Electr Eng 44, 1467–1486 (2020), https://doi.org/10.1007/s40998-020-00341-5.

[64] Ruiping Yang, Xiaojing Lin, Chunhui Tao et al. Based on Network Pharmacology to Explore the Molecular Mechanism of Buzhong Yiqi Decoction for the Treatment of Gastric Cancer, 30 November 2021, PREPRINT (Version 1) available at Research Square. https://doi.org/10.21203/rs.3.rs-1098024/v1.

[65] Zhenpeng Yang, Shuai Lu, Huazhen Tang, Jinxiu Qu, Bing Wang, Yuying Wang, Guofeng Pan, Benqiang Rao, "Molecular Targets and Mechanisms of Hedyotis diffusa-Scutellaria barbata Herb Pair for the Treatment of Colorectal Cancer Based on Network Pharmacology and Molecular Docking", Evidence-Based Complementary and Alternative Medicine, vol. 2022, Article ID 6186662, 15 pages, 2022. https://doi.org/10.1155/2022/6186662.

[66] Liu S, Zhao Y, Duan R, Wu Y, Chen X, Li N. Identification of core genes associated with type 2 diabetes mellitus and gastric cancer by bioinformatics analysis. Ann Transl Med 2022;10(5):247. doi: 10.21037/atm-21-3635.

[67] Zu G, Sun K, Li L, Zu X, Han T, Huang H (2021) Therapeutic Targets and Mechanism of Banxia Xiexin Decoction on Precancerous Lesions of Gastric Cancer: Network Pharmacology. J Clin Trials. S12:003.

[68] T Santh Rani, P Premitha Rajya Lakshmi and Ch Manga Devi, Network pharmacology and molecular docking study of the active ingredients in Saptasaram kashayam for the treatment of Polycystic ovary syndrome, Indian Journal of Biochemistry & Biophysics Vol. 60, February 2023, pp. 108-121, DOI: 10.56042/ijbb.v60i2.70684.

[69] Xavier Robin, Natacha Turck, Alexandre Hainard, et al. (2011) pROC: an open-source package for R and S+ to analyze and compare ROC curves. BMC Bioinformatics, 7, 77. DOI:10.1186/1471-2105-12-77.

[70] Mallik S, Seth S, Si A, Bhadra T, Zhao Z. Optimal ranking and directional signature classification using the integral strategy of multi-objective optimization-based association rule mining of multi-omics data. Front Bioinform. 2023 Jul 27;3:1182176. doi: 10.3389/fbinf.2023.1182176. PMID: 37576714; PMCID: PMC10415913.

4

Dimensionality Reduction, Clustering and Biomarkers Discovery on Single-cell RNA-seq Data

4.1 Introduction

In the medical domains of oncology, microbiology, neurology, reproduction, and immunology, single-cell RNA sequencing (scRNAseq) technology is essential [1]. Using transcriptome single-cell RNA sequencing technique, the genome may be identified. It can also acquire additional multi-omics data to reveal the variations in cell populations and the evolutionary interactions between cells. But, conventional sequencing technique has some drawbacks. For instance, it is unable to evaluate a small number of cells and can only find the average of many cells. Single-cell RNA sequencing technology, which can identify heterogeneity among individual cells, solves the problem of traditional sequencing technology loosing information about cellular heterogeneity. The workflow of single-cell sequencing includes extracting a single cell from a population of cells, analysing cell heterogeneity, molecular mapping, and monitoring immune infiltration and epigenetic alterations. The current studies that specialize in single-cell sequential analysis include classification of various tumour types, investigation of heterogeneity in derived cells, identification of existing and novel cell types, cell representation, cell fate prediction, and cell visualisation [2]. In conducting such a study, single-cell clustering is vital. Cell clustering is necessary for single-cell sequencing analysis in order to identify and examine cluster-specific gene signatures, to reconcile cell type configuration in order to identify the gene signature as differentially expressed, and to simplify the bulk RNA-seq expression data by removing noise. Many scRNA-seq clustering techniques are documented in the scientific literature due to their significance. In contrast, several distance metrics are imported, different techniques for dimension reduction, and different approaches to calculate the number of clusters are used by various clustering methods to increase the accuracy of the results. There are advantages and disadvantages to each clustering technique. Using two or more clustering approaches is advised for cell clustering in order to improve accuracy and thoroughness of analyses. When cluster labels are unavailable, it is crucial to choose the optimum clustering technique. Huh et al. [2] presented a mixture model-based probabilistic framework for single cell clustering in 2020. They did this by combining various clustering techniques, such as aggregate clustering with t-SNE+k-means and ADPclust, an automated method that can evaluate the number of clusters and the centroids of clusters. The performance of clustering for labelling isolated single cells and the correct estimation of the number of clusters was said to have been improved by the technique of authors. Unfortunately, due to high dimensionality, sparse matrix computing, and rare cell detection, their method encounters a number of analytical and technical difficulties while analysing large-scale single-cell data [3]. The curse of dimensionality is specifically caused by a large dimensionality and sparse matrix. For the computational processing of scRNA-seq data, a number of approaches are therefore required, such as quality control, mapping, quantification, dimensionality reduction, grouping, discovering trajectories, and identifying differentially expressed genes. Dimensionality reduction and clustering, which have a significant impact on downstream analysis, are the two most crucial strategies among them.

High dimensionality makes it difficult to identify clusters in single-cell sequencing data. In multiple research domains, dimensionality reduction techniques are required to address this issue as well as other unwanted attributes of high-dimensional space. In a high-dimensional feature space, a set of samples can be naturally grouped together using unsupervised dimensionality reduction techniques. In the Monocle scRNA-seq toolkit [4], the k-means algorithm, a well-known and popular clustering algorithm in data mining [5], is applied. The mechanism of dimension reduction after each split or merge is imported by BackSPIN [6] and pcaReduce [7], which are an extension of hierarchical clustering. This method enhances the identification of small size clusters with better accuracy. The two basic subcategories of dimensionality reduction are feature selection and extraction. A subset of features from the original dataset are chosen during feature selection. A new subspace of features is created through feature extraction, which takes information from the initial set of features. The algorithm known as Principal Component Analysis (PCA) is often used for unsupervised feature extraction. Typically, PCA is used with linear models that transform high dimensional data into low-dimensional space [8].

Dimensionality reduction and clustering have drawn more attention in recent years for the interpretation of single-cell RNA sequencing data. In this chapter, we present an integrated clustering model with dimensionality reduction for the identification of cluster-specific biomarkers in single-cell sequencing data [36]. In a single-cell RNA sequential dataset for a rare intestinal cell type in mice (NCBI accession ID:GSE62270) [9], we used it. Pre-filtering was done on features with fewer number of cells and the cells with fewer number of features. Then, to store the data and analysis together of the dataset, we build a Seurat object [10]. Then, we remove cells that are of poor quality or are dying using quality control measures. Then, in order to normalise the data, we used the global-scaling normalisation technique known as "LogNormalize". Next, we used our dataset to obtain cell-to-cell highly variable features. We next used PCA, a linear dimensionality reduction approach, to project high dimensional data into the optimal low dimensional space. We implemented a graph-based clustering on the cell of the top 10 "significant" principal components (PCs) using the modularity optimization agglomerative clustering algorithm, Louvain [11], after identifying fifty "significant" principal components (PCs) based on strong enrichment of low p-value features. After detecting the cell clusters, using Model-based Analysis of Single-cell Transcriptomics (MAST) [12, 13], an expression feature extraction statistical technique for single-cell sequencing data, we discovered cluster-specific biomarkers . Using *Cytoscape* online tool (version 3.8.0) [14], we further performed degree hub gene network analysis and identified the top five degree markers (*Rps4x*, *Rps18*, *Rpl13a*, *Rps12* and *Rpl18a*). Following that, we used the David 6.8 software tool [15] to perform Gene Set Enrichment Analysis (GESA) to identify enriched KEGG pathways and Gene Ontology (GO) terms such as Biological Process (BP), Cellular Component (CC), and Molecular Function (MF) on the set of clusters indicated markers. In conclusion, the dimensionality reduction and hierarchical agglomerative clustering techniques we presented in this integrated framework effectively identify overlapping biomarkers from single-cell RNA sequencing data as cluster-specific frequent biomarkers.

4.2 Materials and Methods

The steps of our proposed methodology [36] are illustrated as follows and in Fig. 4.1.

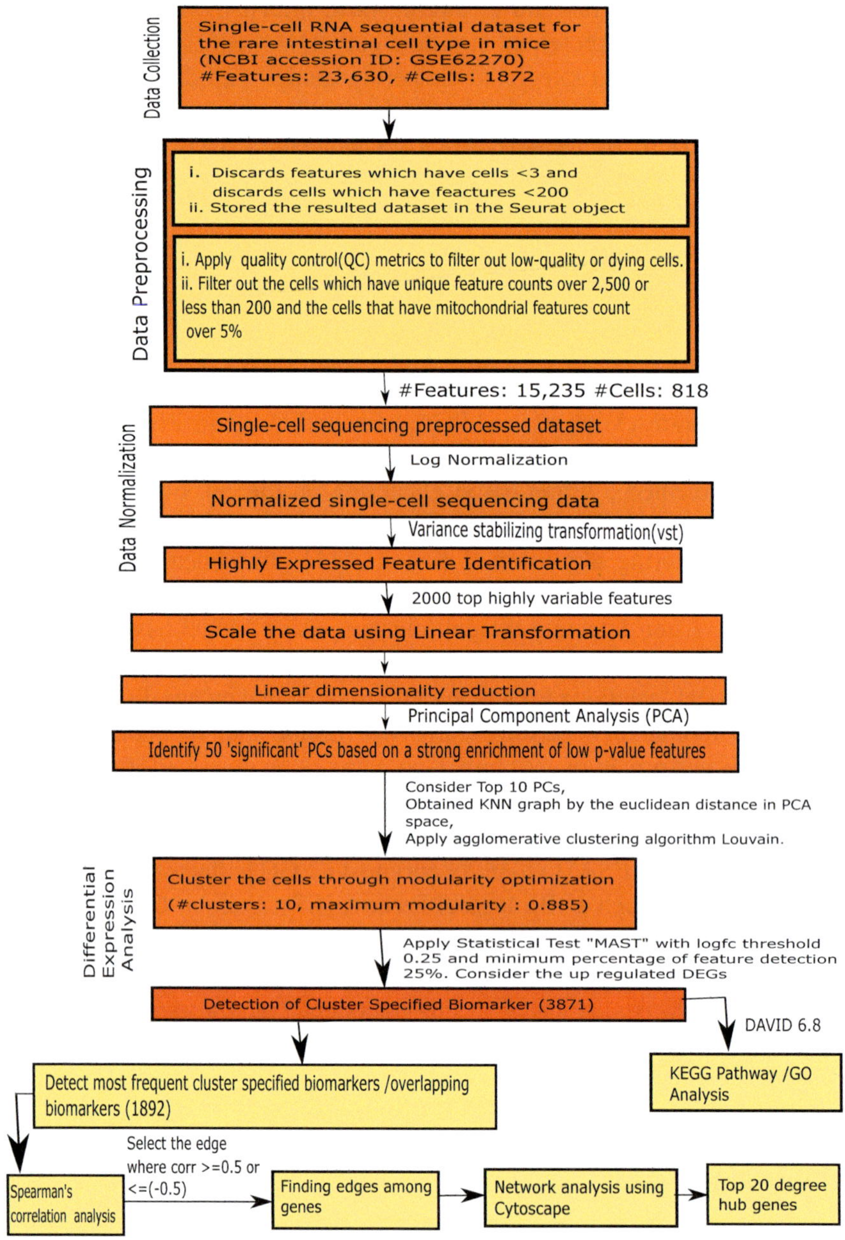

Figure 4.1 Flowchart of the proposed method and follow up analysis.

4.2.1 Data Collection

In this research, we employed a single-cell RNA sequential dataset for the rare intestinal cell type in mice (NCBI accession ID: GSE62270) , which has 23,630 features and 1872 samples (cells) [9].

4.2.2 Preprocessing of Single-cell RNA Sequencing Data

For the purpose of finding cluster-specific frequent biomarkers in single-cell RNA sequencing (scRNAseq) data, we integrated dimensionality reduction approach and clustering algorithm and gave a thorough analysis in this paper. We outline preprocessing steps for a scRNA-seq dataset in the subsections that follow.

4.2.2.1 Data Preprocessing

Data preprocessing is a crucial step before further analysis. Then, we eliminated cells and features that did not meet the required minimums. Then, for a single-cell dataset, we developed a Seurat object to store our data matrix, enabling us to keep both data (such as the input feature sample matrix) and analysis (such as PCA, or clustering outputs), together [10, 16].

4.2.2.2 Compute Quality Control(QC) Metrics and Cell Filtration

QC metrics based on user-defined parameters for cell selection and filtration were explored in this step. Empty cells were initially filtered out. One essential preprocessing method for scRNAseq data is to exclude low-quality or dying cells [17]. In general, cells with few genes are thought to be low-quality or dying cells. An important factor is selecting appropriate thresholds that preserve high-quality cell types without eliminating biologically significant cell types. We specified a threshold range (200–2500) for a number of distinct features in a cell and omitted any that did not fit the requirements. In order to determine the mitochondrial count percentage from the set of features without discarding biologically relevant cell types, we computed mitochondrial quality control measures. We also defined a 5% upper bound barrier for the percentage of mitochondrial count, and we filtered out any cells that were higher than this threshold.

4.2.2.3 Data Normalization

After the cell filtration, data normalisation was carried out using the global-scaling normalization method "LogNormalize," which divides the individual feature counts of each cell by its overall counts, multiplies the result by a scaling factor (10^4), and then applies a natural log-transformation. A Z-scoring metric, in the context of scRNAseq, describes how much the frequency of one cell for a particular feature differs from the mean of the frequencies of all the cells for that feature. Using the log-normalized counts, Z-scores are derived. Let's assume that, the transcription value Tr_{ij} where i= feature (gene) and j = cell. Let, N_j represent the total number of counts for the cell j. We can formulate this normalization procedure by $log_{10}\{(Tr_{ij}/N_j) * 10^4\}$ for each feature i in respect to each cell j.

4.2.3 Highly Variable Features Identification

Using our dataset, we then estimated cell-to-cell highly variable features. Features that are "highly variable" are those that are highly expressed in some cells and lowly expressed in other cells. By highlighting the biological signal, such highly variable genes are crucial for downstream analysis in single-cell datasets [18]. In order to detect highly variable features, we employed a mean-variance relationship model in this study, also known as the variance stabilizing transformation (vst) [19]. There is an inherent mean-variance association in scRNA-seq data. We examined the means and variances of each feature before drawing conclusions about this relationship from the data. After that, we fixed a curve and performed a local fitting of polynomials of degree 2 in order to predict the variance of each feature as a function of its mean. A regularised estimator of variance, where the mean of a feature is already known, defines the global fit. By normalizing the feature count, it is possible to avoid eliminating variations that are higher than expected. Since the expected variance is already known, we formulate the transformation as,

$$y_{ij} = \frac{p_{ij} - \bar{p}_i}{\sigma_i} \tag{4.1}$$

where, y_{ij} is the standardized value of feature i in cell j and p_{ij} denotes the raw value of feature i in cell j, p_i is the mean value of feature i and σ_i is the expected standard deviation of feature i derived from the global mean-variance fit. We fixed the maximum standardised value as $\sqrt{M}$, where M is the total number of cells, to reduce the impact of technical outliers. For each feature, the standard deviations of the values across all cells are calculated. After adjusting for mean expression, this variance represents a measure of single cell dispersion, which aids in the ranking of features.

Following ranking, we identify the top 2,000 features as "highly variable" features with the highest standardized variance.

4.2.4 Linear Transformation and Linear Dimensionality Reduction

For downstream analysis of large dimensional data, linear dimensionality reduction is a crucial step [20]. High-dimensional data are projected onto an optimal low-dimensional environment as input by linear dimensionality reduction techniques. Numerous feature interests can be captured using a variety of methodologies, including covariance, dynamical structure, correlation between data sets, input-output relationships, and margin between data classes. A component of linear dimensionality reduction is feature selection and extraction. We employed PCA, a popular linear dimensionality reduction method that captures covariance as a feature of interest. Data are chosen and extracted using PCA based on increasing variance. As "Principal Component," the features with the highest variance are designated. We standardised the data using a linear transformation before applying PCA. Due to the weight being evenly distributed, the highly expressed features are not dominant. The scaled data was then subjected to PCA algorithm, and the 2000 computed "highly variable" features were taken into account as a feature subset. The PCA approach simultaneously increased interpretability and reduced information loss [21]. We used the JackStraw approach to implement a resampling test, through which we retrieved a subset of the data, in order to measure the dimensionality of a dataset. On the subset, PCA and a random permutation (with the default value of 1%) were performed. We run the PCA algorithm again and continue the process to create a null distribution of feature scores. Based on a notable enrichment of low p-value features, we found fifty "significant" principal components (PCs) [22].

4.2.5 Cell Cluster

During this step, we used graph-based clustering to analyse the data and took the first 10 PCs into account. First, using the Euclidean distance in PCA space, we were able to create a K-nearest neighbour (KNN) graph. By the Jaccard similarity, which is determined by shared overlap between cells, we calculated the edge weight between two cells. Suppose, there are two cells C1 and C2, Jaccard similarity is defined by

$$J(C1, C2) = \frac{|C1 \cap C2|}{|C1 \cup C2|} \times 100 \tag{4.2}$$

Cells having the greatest degree of similarity among their features have high Jaccard similarity scores.

The Louvain algorithm [11] was then used for cell clustering, which is a modularity optimization agglomerative clustering technique. In 2006, M.E.J. Newman developed modularity, a significant quality control step for community discovery (clustering) [23]. By comparing the degree of interactions occurring within and between clusters, one can determine the modularity of a graph partition. In the range [-1,1], the modularity value is located. The value of modularity (Q) is formulated by

$$Q = \frac{1}{2e} \sum_{i,j} \left[Adj_{ij} - \frac{w_i w_j}{2e} \right] \delta(Cl_i, Cl_j) \tag{4.3}$$

where, Adj_{ij} denotes weight of the edge between i and j of our KNN graph, i.e. adjacency matrix, $w_i = \sum_j Adj_{ij}$ is the total weights of the edges attached to vertex (here, cell) i, Cl_i is the cluster to which cell i is assigned, δ-function is derived as $\delta(x.y)$= 1 when $x = y$, otherwise it is 0 and $e = \frac{1}{2} \sum_{ij} Adj_{ij}$ [11].

Maximizing the average modularity of computed clusters is the goal of modularity optimization. The Louvain Algorithm, introduced by Blondel et al. in 2008 [11], is an algorithm for modularity optimization. The primary workflow of this method is divided into two steps that are iteratively repeated [24]. They increased local modularity in phase 1 by relocating each node to a community near its neighbors. In more detail, the authors identified the neighbors j of each node i and assessed the gain in modularity that would result from transferring each node from its assigned community to the community (or cluster) of neighboring node j (Cl_j). The node i is located in that community (cluster) which maximizes modularity gain. Gain in modularity ought to be positive. The Modularity gain ΔQ is computed by

$$\Delta Q = \left[\frac{\sum_{Cl_j, Cl_j} + \sum_{i, Cl_j}}{2e} - \left(\frac{\sum_{Cl_j} + \sum_i}{2e} \right)^2 \right] - \left[\frac{\sum_{Cl_j, Cl_j}}{2e} - \left(\frac{\sum_{Cl_j}}{2e} \right)^2 - \left(\frac{\sum_i}{2e} \right)^2 \right] \tag{4.4}$$

where, $\sum_{Cl_j,Cl_j}$ denotes the weights sum of the links which lie in Cl_j, $\sum_{i,Cl_j}$ refers the weights sum of the links from node i to nodes in Cl_j, $\sum_{Cl_j}$ denotes the sum of the weights of the links incident to nodes in Cl_j, $\sum_i$ refers the weights sum of the links incident to node i and e denotes the sum of the weights of all the links in the graph.

The next phase is initiated when the modularity gain reaches its local maximum. The authors formed a new network in the second phase by using the communities they had identified in the first phase as nodes and taking into account the fact that the weights of the links between the new nodes are simply the weight sum of the links between the nodes of the corresponding two communities. The self-loops in the new network are links between nodes belonging to the same community. The first two phases are continued until the modularity gain is constant and the maximum modularity is preserved. In our work, we used the R tool FindCluster() to cluster the cells using the community detection algorithm. The parameter 'resolution' was used to set the granularity of the downstream analysis, which determines the number of clusters. When the value of resolution parameter increased, more clusters were detected. For single-cell datasets with about 3000 cells, setting this parameter to the range 0.4-1.2 often yields significant results, according to the benchmark. Uniform Manifold Approximation and Projection (UMAP), a non-linear dimensional reduction technique, was also used to show the identical cells of graph-based clusters in low-dimensional space while having the same number of PCs as we discovered during cluster analysis. [25].

4.2.6 Finding the Cluster-Specific Biomarkers

We explain how we identified the genes from each cluster that were differently expressed i.e. identifying the biomarkers that were specific to each cluster. Model-based Analysis of Single-cell Transcriptomics, a commonly used statistical approach for single-cell sequencing data, was used to extract differentially expressed features (markers) (MAST) [12]. To detect differentially expressed genes from each cluster, we repeated the technique using all remaining clusters as the Control Group and each cluster as the Experimental Group. A two-part generalised regression model was used by MAST to model the gene expression matrix. It built a logistic regression model to represent the rate of gene expression, and then utilized a Gaussian distribution model at the expression level to apply conditions to cells that were expressing the gene [12]. In bimodal expression distributions where expression is either significantly non-zero or undetectable, the MAST model is quite useful [13]. Frank et al. (2015) developed the MAST model [12]. The authors denoted $Y = [Y_{ig}]$ as the rate of expression and the level of expression for an independent gene g and cell i. They used an indicator $Z = [z_{ig}]$ to indicate whether the gene g is expressed in the cell i or not (i.e., $z_{ig} = 0$ if $y_{ig} = 0$ and $z_{ig} = 1$ if $y_{ig} > 0$). The Authors formulated a logistic regression model for the discrete variable Z and a Gaussian linear model for the continuous variable $(Y|Z = 1)$ as follow [12, 13]:

$$logit(P_r(Z_{ig} = 1)) = X_i \beta_g^D \tag{4.5}$$

$$P_r = (Y_{ig} = y|Z_{ig} = 1) = N(X_i \beta_g^C, \sigma_g^2) \tag{4.6}$$

X_i is the design matrix. Fraction of genes being expressed and detectable in each cell, is termed as cellular detection rate (CDR). The CDR for cell i is formulated as:

$$CDR_i = \frac{1}{N} \sum_{g=1}^{N} z_{ig} \tag{4.7}$$

In both the discrete and continuous models, CDR variability is indicated by a covariate variable (a column in the design matrix X_i). CDR covariate is significant because it is the discrete equivalent of global normalisation, which by reducing background feature correlation helps identify true gene co-expression. The total number of genes in a cell is called N. An empirical Bayesian framework is used to fit the parameters of this mode, which enhances the inference for genes with low expression. Tests for differential expression are performed using the likelihood ratio. β is the likelihood estimator, σ denotes variance.

In addition to the MAST model, we used the logarithm of fold change with base 2 ($log_2 FC$) threshold to determine how much a feature should be differentially expressed. We fixed this threshold at 0.25 and selected up-regulated differentially expressed genes from all clusters as markers. Another parameter, min.pct, was simultaneously set to 0.25, holding a minimum percentage (i.e. 25%) of feature detection in either of the two groups, Control group and Experimental group. Based on p-value and False Discovery Rate (FDR) corrected adjusted p-value, we discovered cluster-specific biomarkers. Finding frequent cluster markers i.e. overlapping markers was our main objective.

4.2.7 Hub Gene Finding

The following phase was using Spearman's correlation analysis on the most frequent and cluster-specific biomarkers discovered by our technique. Obtaining the active edges between genes with correlation values ≥ 0.5 or ≤ -0.5 is the goal of this step. We used *Cytoscape* online tool (version 3.8.0) [14] to perform degree centrality analysis and calculate degree scores for each marker after getting the set of active edges. The top 20 markers (hubs) were ranked by degree.

4.2.8 Gene Set Enrichment Analysis

A list of signature genes is evaluated for potential function, biological significance, and disease relevance using the Gene Set Enrichment Analysis (GESA) method. After finding cluster-specific biomarkers and differentially expressed genes, we employed KEGG pathways and Gene Ontology (GO) annotations (which are divided into three domains: Biological Process (BP), Cellular Component (CC), and Molecular Function (MF)) using DAVID 6.8 software [15]. The total number of genes in each KEGG pathway and Gene Ontology (GO) term, as well as the enriched adjusted p-value and FDR, were obtained. If the FDR was 0.05 or less, we retained the KEGG pathways or GO terms.

4.3 Result and Discussion

In this work, we employed a single-cell RNA sequential dataset for the rare intestinal cell type in mice (GEO ID: GSE62270), which comprises 23,630 features and 1872 samples [9]. Our scRNA-seq data were prepared for additional analysis. Before storing the resulting dataset as a Seurat object, we first removed features with fewer than three cells and cells with fewer than 200 features. The final dataset had 1644 samples and 15,235 features. To remove poor-quality or dying cells that cause mitochondrial pollution, we used various quality control (QC) measures. Also, we computed QC metrics using the PercentageFeatureSet function, which can determine the proportion of mitochondrial feature count from a batch of all kinds of features. Here, we defined the list of mitochondrial genes as all genes beginning with 'MT'. We eliminated cells with unique feature counts of more than 2,500 or less than 200 as well as those with mitochondrial feature counts of more than 5%. Our dataset contains 818 cells (samples) with 15,235 features after these preprocessing steps. Fig. 4.2 visualizes QC metrics as a violin plot and Fig. 4.3 shows the visualization of feature-feature relationships. After eliminating irrelevant cells, we normalized our dataset using the log normalization approach and identified highly variable features using the "vst"

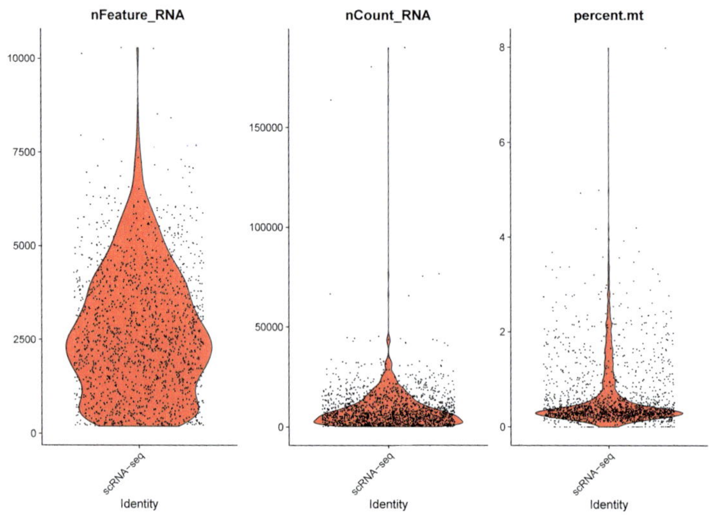

Figure 4.2 Visualize QC metrics as a violin plot.

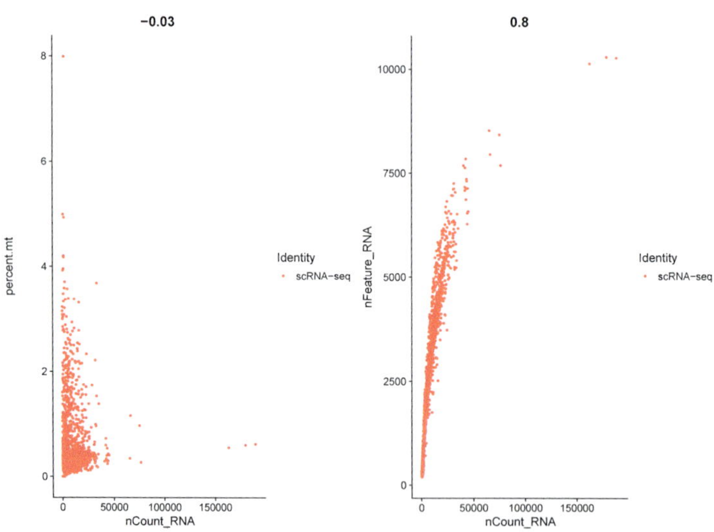

Figure 4.3 FeatureScatter plot to visualize feature-feature relationships.

feature selection method. To carry out further downstream analysis, the 2000 highly variable features were retrieved. Fig. 4.4 displays 2000 highly variable features with some feature labels.

The next stage involves doing PCA, a linear dimensional reduction approach, where we input 2000 highly variable features. As a typical pre-processing step for dimensions reduction technique, we started by applying linear transformation. Equal weights are assigned in downstream analysis through linear transformation, reducing the possibility of dominance by highly expressed traits. The scaled data was then subjected to PCA. Fifty principal components (PC) were obtained, and they included both positive and negative features. Fig. 4.5 refers to Dim Plot of the first principal component vs the second principal component. DimHeatMap for principal components is shown in Fig. 4.6 (Fig. 4.3 and Fig. 4.3 represent DimHeatMap for PC1 and for PC1-PC15 respectively). The term "Significant" PCs refers to PCs with substantial feature enrichment and low p-values. Fig. 4.7 shows fifteen 'Significant' PCs with their respective p-values.

Then, using the Euclidean distance in PCA space, we built a K-nearest neighbour (KNN) graph and used the Jaccard similarity to calculate the edge weights between any two cells. We used the top 10 PCs to determine the

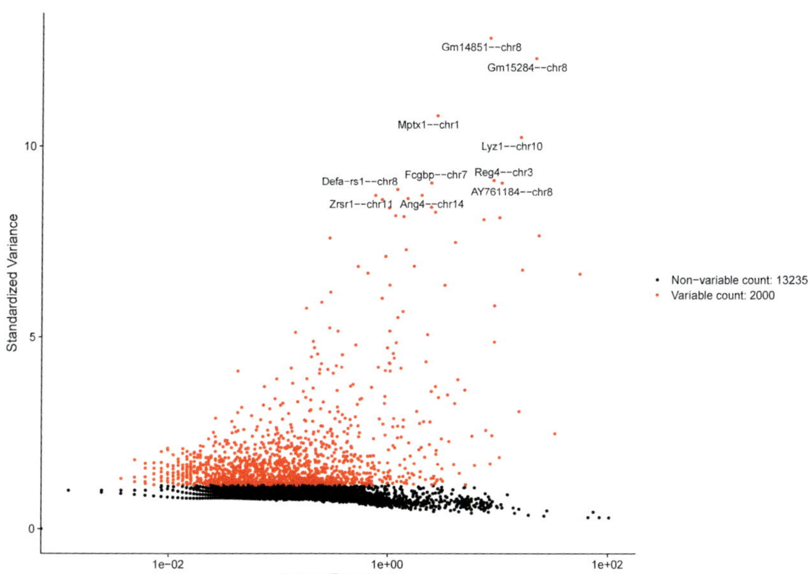

Figure 4.4 Variable features with labels.

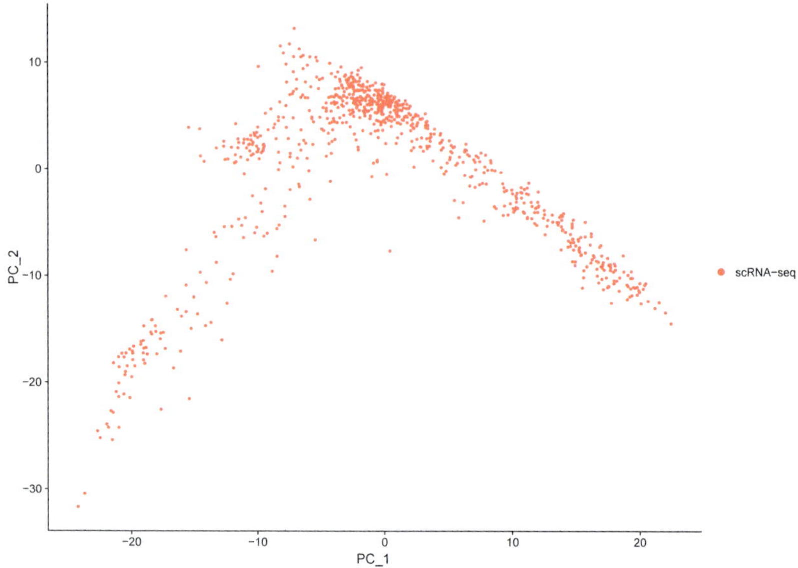

Figure 4.5 DimPlot of two principal components (PC1 Vs PC2).

dimensionality of the dataset. There were 818 nodes and 20,511 edges in our KNN graph. We used a modularity optimization strategy for cell clustering (Louvain algorithm). Resolution is a parameter used by this clustering procedure to control the 'granularity' of the downstream clustering. A bigger number of clusters result from raising the value of resolution parameter. For single-cell datasets with about 3000 cells, setting this parameter to the range 0.4-1.2 often yields significant results, according to benchmark. With the value set at 0.5 for our analysis, we got 10 clusters with a maximum modularity of 0.8851. The similar cells of the graph-based clusters were then visualised in low-dimensional space using a non-linear dimensional reduction approach called UMAP (referred Fig. 4.8). We took into account the identical PCs that the cluster analysis detected.

Finding cluster-specific biomarkers, also known as differentially expressed features, is the last phase in our process. We treated each cluster as a separate group and the other clusters as a different group. We used the statistical test "MAST" to find features that were differentially expressed between two groups of cells. "MAST" employs a barrier model designed for scRNA-seq data. In this process, we used up-regulated differentially

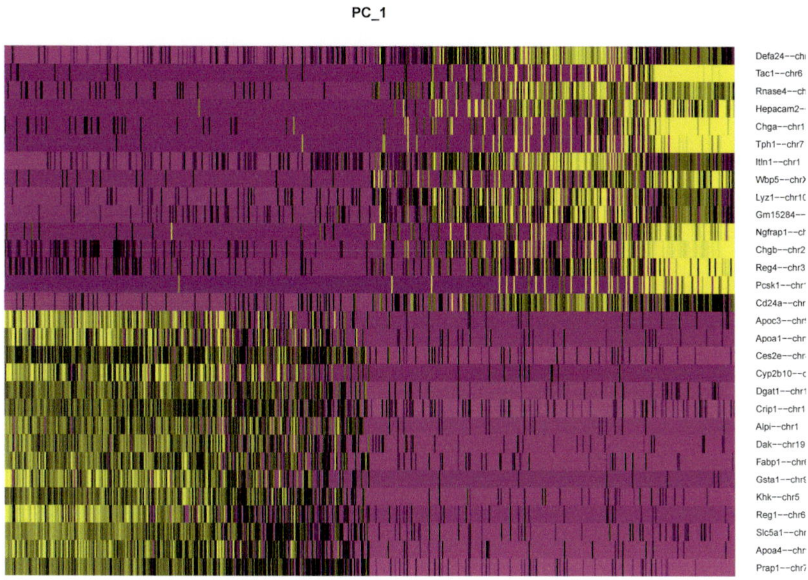

(a) DimHeatMap for first principal component

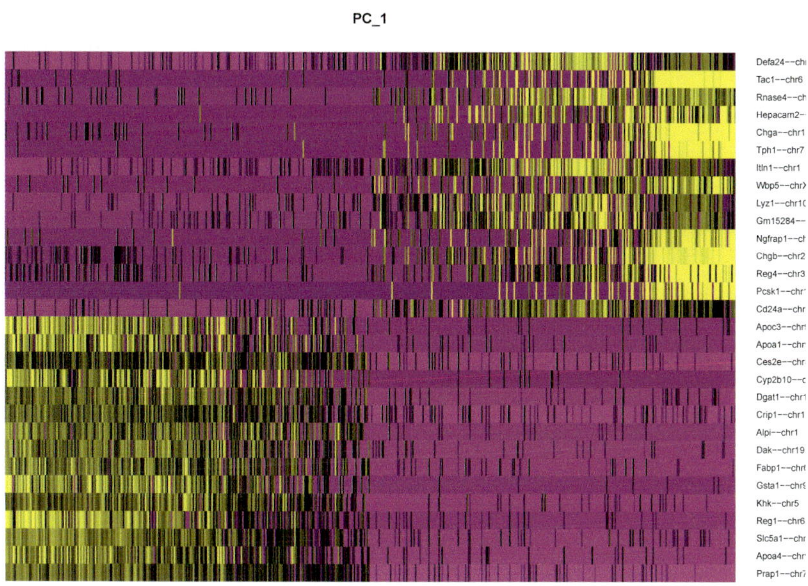

(b) DimHeatMap for fifteen principal components

Figure 4.6 DimHeatMap for principal components

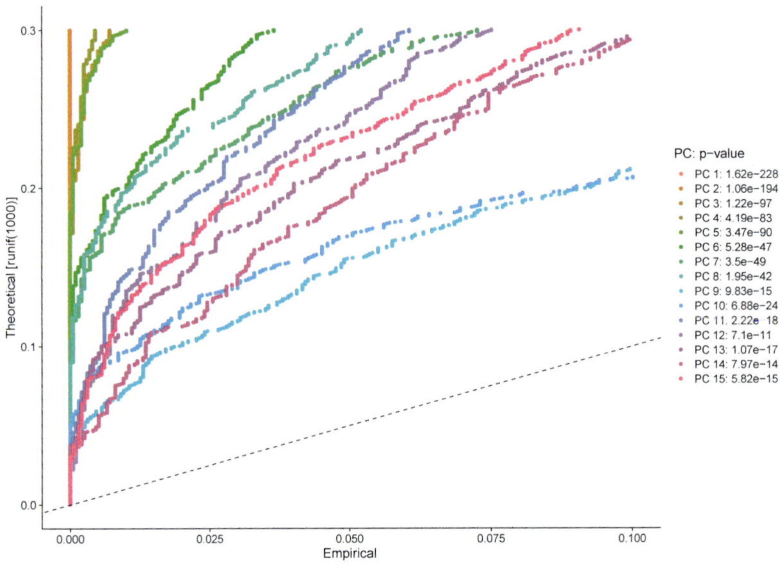

Figure 4.7 Visualization of strong enrichment of features with low p-values.

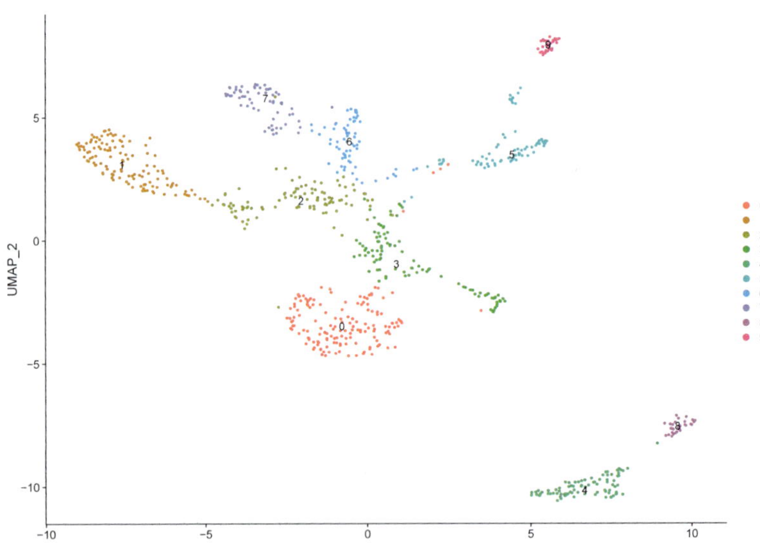

Figure 4.8 Visualization of clusters.

expressed genes from all clusters as markers and adjusted the log_2FC threshold, which measures how much a feature is differently expressed, to 0.25. The minimal percentage of a feature detection in any of the two groups, known as min.pct, is set to 0.25. We discovered 6394 cluster-specific markers with their associated clusters, of which 3871 markers are unique. Several of the overlapping markers were found in multiple clusters. The term "frequent marker" refers to these markers. In our analysis, we found 1892 frequent biomarkers. Table 4.1 presents the thirty highest frequent biomarkers together with their frequency and adjusted p-values based on Bonferroni correction utilising all the genes in the dataset. This adjusted p-value is also known as

Table 4.1 Top 30 Cluster Specified Frequent Biomarkers

Marker Name	Frequency	Specified Clusters	FDR
Atp5j2	$\frac{5}{10}$	Group1: Cluster0 Vs Group2: Rest all clusters	1.82×10^{-06}
		Group1: Cluster1 Vs Group2: Rest all clusters	7.12×10^{-03}
		Group1: Cluster3 Vs Group2: Rest all clusters	2.49×10^{-07}
		Group1: Cluster6 Vs Group2: Rest all clusters	1.00×10^{-00}
		Group1: Cluster7 Vs Group2: Rest all clusters	1.00×10^{-00}
ERCC_00009	$\frac{5}{10}$	Group1: Cluster2 Vs Group2: Rest all clusters	8.69×10^{-28}
		Group1: Cluster3 Vs Group2: Rest all clusters	2.08×10^{-02}
		Group1: Cluster6 Vs Group2: Rest all clusters	4.82×10^{-05}
		Group1: Cluster7 Vs Group2: Rest all clusters	3.24×10^{-01}
		Group1: Cluster8 Vs Group2: Rest all clusters	1.14×10^{-04}
Nedd4	$\frac{5}{10}$	Group1: Cluster0 Vs Group2: Rest all clusters	1.07×10^{-8}
		Group1: Cluster3 Vs Group2: Rest all clusters	3.45×10^{-07}
		Group1: Cluster4 Vs Group2: Rest all clusters	2.29×10^{-04}
		Group1: Cluster5 Vs Group2: Rest all clusters	6.97×10^{-01}
		Group1: Cluster9 Vs Group2: Rest all clusters	4.97×10^{-03}
Abhd17a	$\frac{4}{10}$	Group1: Cluster1 Vs Group2: Rest all clusters	6.52×10^{-02}
		Group1: Cluster2 Vs Group2: Rest all clusters	4.45×10^{-05}
		Group1: Cluster6 Vs Group2: Rest all clusters	1.00×10^{-00}
		Group1: Cluster7 Vs Group2: Rest all clusters	8.31×10^{-02}
Actn1	$\frac{4}{10}$	Group1: Cluster0 Vs Group2: Rest all clusters	1.29×10^{-02}
		Group1: Cluster4 Vs Group2: Rest all clusters	1.50×10^{-06}
		Group1: Cluster5 Vs Group2: Rest all clusters	8.56×10^{-05}
		Group1: Cluster9 Vs Group2: Rest all clusters	8.12×10^{-05}
Arpc1b	$\frac{4}{10}$	Group1: Cluster0 Vs Group2: Rest all clusters	1.04×10^{-05}
		Group1: Cluster1 Vs Group2: Rest all clusters	3.00×10^{-04}
		Group1: Cluster3 Vs Group2: Rest all clusters	1.69×10^{-05}
		Group1: Cluster8 Vs Group2: Rest all clusters	4.63×10^{-03}
Atp5b	$\frac{4}{10}$	Group1: Cluster0 Vs Group2: Rest all clusters	7.59×10^{-11}
		Group1: Cluster3 Vs Group2: Rest all clusters	2.34×10^{-05}
		Group1: Cluster6 Vs Group2: Rest all clusters	9.96×10^{-05}
		Group1: Cluster7 Vs Group2: Rest all clusters	1.00×10^{-04}

Table 4.1 *Continued*

Marker Name	Frequency	Specified Clusters	FDR
B2m	$\frac{4}{10}$	Group1: Cluster0 Vs Group2: Rest all clusters	6.49×10^{-15}
		Group1: Cluster1 Vs Group2: Rest all clusters	1.14×10^{-15}
		Group1: Cluster3 Vs Group2: Rest all clusters	8.88×10^{-11}
		Group1: Cluster9 Vs Group2: Rest all clusters	5.56×10^{-03}
B4galnt2	$\frac{4}{10}$	Group1: Cluster0 Vs Group2: Rest all clusters	9.46×10^{-16}
		Group1: Cluster1 Vs Group2: Rest all clusters	9.37×10^{-04}
		Group1: Cluster3 Vs Group2: Rest all clusters	1.41×10^{-02}
		Group1: Cluster4 Vs Group2: Rest all clusters	5.64×10^{-07}
Calm1	$\frac{4}{10}$	Group1: Cluster1 Vs Group2: Rest all clusters	6.53×10^{-18}
		Group1: Cluster3 Vs Group2: Rest all clusters	8.54×10^{-04}
		Group1: Cluster5 Vs Group2: Rest all clusters	1.28×10^{-05}
		Group1: Cluster9 Vs Group2: Rest all clusters	1.65×10^{-01}
Cox5b	$\frac{4}{10}$	Group1: Cluster0 Vs Group2: Rest all clusters	1.72×10^{-14}
		Group1: Cluster3 Vs Group2: Rest all clusters	9.78×10^{-07}
		Group1: Cluster6 Vs Group2: Rest all clusters	5.75×10^{-01}
		Group1: Cluster7 Vs Group2: Rest all clusters	1.18×10^{-03}
Dhcr24	$\frac{4}{10}$	Group1: Cluster1 Vs Group2: Rest all clusters	3.35×10^{-10}
		Group1: Cluster6 Vs Group2: Rest all clusters	3.83×10^{-03}
		Group1: Cluster7 Vs Group2: Rest all clusters	1.15×10^{-04}
		Group1: Cluster8 Vs Group2: Rest all clusters	4.44×10^{-01}
Dpysl2	$\frac{4}{10}$	Group1: Cluster4 Vs Group2: Rest all clusters	3.26×10^{-05}
		Group1: Cluster5 Vs Group2: Rest all clusters	2.29×10^{-05}
		Group1: Cluster8 Vs Group2: Rest all clusters	8.48×10^{-08}
		Group1: Cluster9 Vs Group2: Rest all clusters	6.19×10^{-01}
Dst	$\frac{4}{10}$	Group1: Cluster1 Vs Group2: Rest all clusters	1.34×10^{-06}
		Group1: Cluster5 Vs Group2: Rest all clusters	8.44×10^{-03}
		Group1: Cluster8 Vs Group2: Rest all clusters	4.38×10^{-03}
		Group1: Cluster9 Vs Group2: Rest all clusters	3.29×10^{-02}
Eef1a1	$\frac{4}{10}$	Group1: Cluster0 Vs Group2: Rest all clusters	2.84×10^{-36}
		Group1: Cluster4 Vs Group2: Rest all clusters	1.85×10^{-03}
		Group1: Cluster5 Vs Group2: Rest all clusters	4.43×10^{-02}
		Group1: Cluster9 Vs Group2: Rest all clusters	6.96×10^{-01}
ERCC_00003	$\frac{4}{10}$	Group1: Cluster1 Vs Group2: Rest all clusters	8.53×10^{-07}
		Group1: Cluster2 Vs Group2: Rest all clusters	6.58×10^{-10}
		Group1: Cluster3 Vs Group2: Rest all clusters	3.11×10^{-07}
		Group1: Cluster8 Vs Group2: Rest all clusters	4.55×10^{-19}
ERCC_00043	$\frac{4}{10}$	Group1: Cluster1 Vs Group2: Rest all clusters	3.14×10^{-09}
		Group1: Cluster2 Vs Group2: Rest all clusters	4.50×10^{-10}
		Group1: Cluster3 Vs Group2: Rest all clusters	5.90×10^{-03}
		Group1: Cluster8 Vs Group2: Rest all clusters	2.37×10^{-27}
ERCC_0007	$\frac{4}{10}$	Group1: Cluster1 Vs Group2: Rest all clusters	1.04×10^{-06}
		Group1: Cluster2 Vs Group2: Rest all clusters	7.88×10^{-02}
		Group1: Cluster3 Vs Group2: Rest all clusters	9.58×10^{-12}
		Group1: Cluster8 Vs Group2: Rest all clusters	6.17×10^{-18}
Fdps	$\frac{4}{10}$	Group1: Cluster1 Vs Group2: Rest all clusters	4.45×10^{-04}
		Group1: Cluster2 Vs Group2: Rest all clusters	6.28×10^{-01}
		Group1: Cluster6 Vs Group2: Rest all clusters	1.91×10^{-08}
		Group1: Cluster7 Vs Group2: Rest all clusters	8.40×10^{-05}
Glud1	$\frac{4}{10}$	Group1: Cluster0 Vs Group2: Rest all clusters	9.60×10^{-05}
		Group1: Cluster3 Vs Group2: Rest all clusters	3.21×10^{-01}
		Group1: Cluster5 Vs Group2: Rest all clusters	2.70×10^{-11}
		Group1: Cluster9 Vs Group2: Rest all clusters	2.21×10^{-07}
Hsp90b1	$\frac{4}{10}$	Group1: Cluster0 Vs Group2: Rest all clusters	2.48×10^{-03}
		Group1: Cluster4 Vs Group2: Rest all clusters	8.20×10^{-04}
		Group1: Cluster5 Vs Group2: Rest all clusters	7.20×10^{-01}
		Group1: Cluster9 Vs Group2: Rest all clusters	3.15×10^{-06}

Table 4.1 *Continued*

Marker Name	Frequency	Specified Clusters	FDR
Malat1	$\frac{4}{10}$	Group1: Cluster1 Vs Group2: Rest all clusters	2.38×10^{-12}
		Group1: Cluster2 Vs Group2: Rest all clusters	1.91×10^{-02}
		Group1: Cluster5 Vs Group2: Rest all clusters	2.98×10^{-06}
		Group1: Cluster9 Vs Group2: Rest all clusters	2.76×10^{-07}
Rpl17	$\frac{4}{10}$	Group1: Cluster0 Vs Group2: Rest all clusters	1.34×10^{-39}
		Group1: Cluster3 Vs Group2: Rest all clusters	1.00×10^{-00}
		Group1: Cluster4 Vs Group2: Rest all clusters	1.16×10^{-02}
		Group1: Cluster9 Vs Group2: Rest all clusters	8.70×10^{-01}
Rpl4	$\frac{4}{10}$	Group1: Cluster0 Vs Group2: Rest all clusters	3.65×10^{-35}
		Group1: Cluster3 Vs Group2: Rest all clusters	7.05×10^{-02}
		Group1: Cluster4 Vs Group2: Rest all clusters	9.36×10^{-03}
		Group1: Cluster9 Vs Group2: Rest all clusters	1.00×10^{-00}
Rpl7	$\frac{4}{10}$	Group1: Cluster0 Vs Group2: Rest all clusters	5.36×10^{-29}
		Group1: Cluster4 Vs Group2: Rest all clusters	3.64×10^{-01}
		Group1: Cluster6 Vs Group2: Rest all clusters	3.68×10^{-01}
		Group1: Cluster9 Vs Group2: Rest all clusters	1.00×10^{-00}
Rps6	$\frac{4}{10}$	Group1: Cluster0 Vs Group2: Rest all clusters	1.39×10^{-29}
		Group1: Cluster4 Vs Group2: Rest all clusters	5.02×10^{-03}
		Group1: Cluster6 Vs Group2: Rest all clusters	1.00×10^{-00}
		Group1: Cluster9 Vs Group2: Rest all clusters	3.52×10^{-01}
Sox9	$\frac{4}{10}$	Group1: Cluster0 Vs Group2: Rest all clusters	3.56×10^{-06}
		Group1: Cluster4 Vs Group2: Rest all clusters	1.63×10^{-10}
		Group1: Cluster5 Vs Group2: Rest all clusters	1.81×10^{-01}
		Group1: Cluster9 Vs Group2: Rest all clusters	1.19×10^{-03}
Tkt	$\frac{4}{10}$	Group1: Cluster1 Vs Group2: Rest all clusters	1.24×10^{-05}
		Group1: Cluster2 Vs Group2: Rest all clusters	5.81×10^{-05}
		Group1: Cluster6 Vs Group2: Rest all clusters	1.68×10^{-02}
		Group1: Cluster7 Vs Group2: Rest all clusters	2.44×10^{-01}
Tm9sf3	$\frac{4}{10}$	Group1: Cluster0 Vs Group2: Rest all clusters	1.42×10^{-03}
		Group1: Cluster4 Vs Group2: Rest all clusters	2.90×10^{-14}
		Group1: Cluster5 Vs Group2: Rest all clusters	1.98×10^{-02}
		Group1: Cluster9 Vs Group2: Rest all clusters	1.77×10^{-03}
Ywhae	$\frac{4}{10}$	Group1: Cluster0 Vs Group2: Rest all clusters	3.64×10^{-09}
		Group1: Cluster3 Vs Group2: Rest all clusters	3.75×10^{-08}
		Group1: Cluster5 Vs Group2: Rest all clusters	9.12×10^{-05}
		Group1: Cluster9 Vs Group2: Rest all clusters	2.26×10^{-02}

an adjusted false discovery rate (FDR) p-value. As a supplementary file, we included a list of all frequently encountered biomarkers.

Furthermore, we used Spearman's correlation analysis to find edges between genes with correlation values greater than or equal to 0.5 (strongly positively associated) or less than or equal to (-0.5)(highly negatively correlated) on our cluster-specific most frequent markers or overlapped markers. Then, utilising *Cytoscape* [14], we carried out degree centrality hub gene network analysis. In our analysis, five genes with the highest degree values were *Rps4x*, *Rps18*, *Rpl13a*, *Rps12* and *Rpl18a*, see Table 4.2. We illustrated a detailed hub gene network structure in Fig. 4.9.

Most of the top hub genes identified by our technique were found to be highly expressed markers or transcripts in the exceptional nature of cell detection in the related literature study. In the study of An et al. [26], the

Table 4.2 Top 20 hub genes ranked by degree centrality.

Gene Symbol	Degree	Average Shortest Path Length	Betweenness Centrality	Closeness Centrality	Clustering Coefficient
Rps4x	32	1.800	0.056	0.556	0.536
Rps18	32	1.861	0.065	0.537	0.566
Rpl13a	31	1.877	0.021	0.533	0.596
Rps12	29	1.892	0.018	0.528	0.640
Rpl18a	29	1.923	0.012	0.520	0.662
Gnb2l1	29	1.923	0.036	0.520	0.589
Rps8	28	1.862	0.034	0.537	0.600
Rps7	28	1.661	0.099	0.602	0.587
Rpl23	28	1.646	0.170	0.607	0.582
Rpl39	27	1.877	0.032	0.533	0.587
Rps17	25	1.985	0.008	0.504	0.740
Rps6	24	1.923	0.018	0.520	0.677
Rps9	24	2.000	0.011	0.500	0.688
Rpl3	23	2.015	0.005	0.496	0.794
Rpl7	23	1.969	0.010	0.508	0.735
Eef1a1	22	1.785	0.142	0.560	0.420
Gm13826	22	2.046	0.004	0.489	0.770
Rplp1	22	2.323	0.007	0.430	0.675
Gm6402	21	2.000	0.018	0.500	0.628
Rplp0	21	2.338	0.008	0.428	0.652

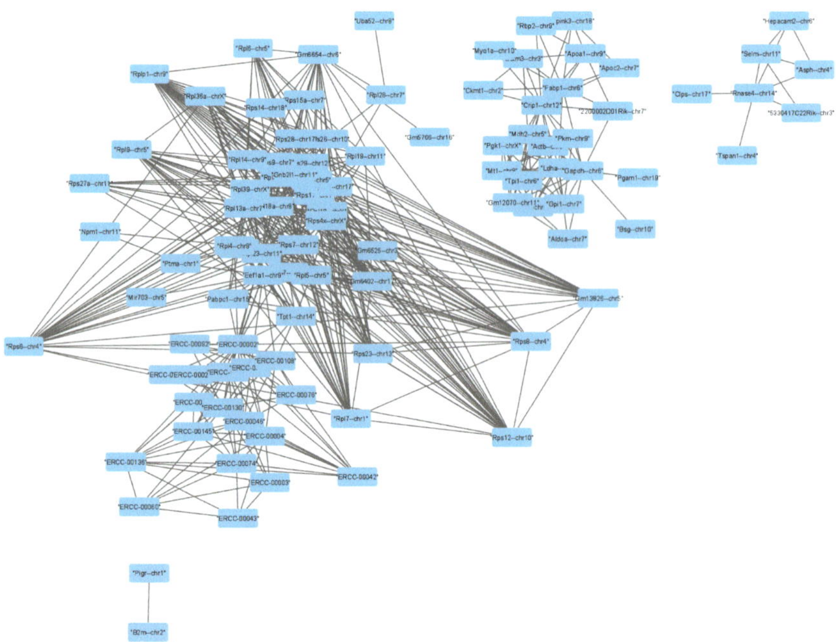

Figure 4.9 Visualization of Hub gene network of strongly correlated frequent markers.

Rps4x marker was regarded as a highly expressed transcript, and Bradley et al. [27] found that it was crucial to identifying the special nature of the X chromosome. *Rps18* was found by Matarin et al. [28] to be the variant that minimises pairwise gene expression variation in the hippocampus tissues of different mice. Wright et al. [29] discovered the *Rpl13a* marker as a housekeeping gene that is highly expressed in all types of cells, while Fa et al. [30] discovered it as an up-regulated marker to recover the rare CD 34+ cells. In the investigation by Wisdom et al. [31], the marker *Rps12* was discovered as the antibody boosted the expression of proliferation genes. *Rpl18a* was identified by Basak et al. [32] as a cluster-specific overlapping marker that is present in three clusters.

Additionally, we used David 6.8 software and 3871 cluster-specific markers to perform Gene Set Enrichment Analysis [15]. The KEGG pathways and Gene Ontology (GO) terms [Biological Process (BP), Cellular Component (CC), and Molecular Function (MF)] along with the number of genes in each pathway or GO-term, an enriched Bonferroni corrected p-value, and FDR were obtained using the DAVID database on our cluster-specific markers. We import our input dataset using the recommended format of DAVID 6.8 software, which calls for a list of gene names in a single column, $OFFICIAL\_GENE\_SYMBOL$ as the Identifier, and Mus musculus as the Species. There are several significant paths and GO-terms listed in below , and the Tables 4.3-4.6 provide more information. The KEGG pathway for *mmu03010:Ribosome* had the lowest FDR value (5.17×10^{-40}). 112 genes in total had enriched Bonferroni adjusted p-values of 7.54×10^{-40} linked with this pathway. The remaining ten major KEGG pathways are included in Table 4.3. As a supplementary file, we supplied a list of all KEGG pathways. In a similar vein, the term *GO:0006412 translation*, with an FDR value

Table 4.3 Top significant KEGG Pathways (FDR sorted)

KEGG pathway name	#genes	Enriched Adjusted p-value	FDR
mmu03010:Ribosome	112	7.54×10^{-40}	5.17×10^{-40}
mmu01100:Metabolic pathways	477	3.97×10^{-33}	1.36×10^{-33}
mmu01130:Biosynthesis of antibiotics	124	5.34×10^{-25}	1.22×10^{-25}
mmu00190:Oxidative phosphorylation	89	3.77×10^{-22}	2.58×10^{-22}
mmu05016:Huntington's disease	111	1.50×10^{-20}	2.05×10^{-21}
mmu05012:Parkinson's disease	90	2.79×10^{-19}	3.19×10^{-20}
mmu05010:Alzheimer's disease	100	1.19×10^{-18}	1.16×10^{-19}
mmu04932:Non-alcoholic fatty liver disease (NAFLD)	86	1.14×10^{-14}	9.74×10^{-16}
mmu01200:Carbon metabolism	70	1.50×10^{-14}	1.14×10^{-15}
mmu03040:Spliceosome	74	9.63×10^{-13}	6.57×10^{-14}

Table 4.4 Top significant GO-BP term enriched (FDR sorted)

GO-BP Term Name	#genes	Enriched Adjusted p-value	FDR
GO:0006412 translation	195	1.04×10^{-39}	1.02×10^{-39}
GO:0006810 transport	529	6.16×10^{-28}	3.02×10^{-28}
GO:0055114 oxidation-reduction process	241	2.04×10^{-23}	6.69×10^{-24}
GO:0098609 cell-cell adhesion	93	1.04×10^{-17}	2.55×10^{-18}
GO:0015031 protein transport	198	4.07×10^{-15}	7.99×10^{-16}
GO:0006397 mRNA processing	127	6.55×10^{-15}	9.33×10^{-16}
GO:0008380 RNA splicing	104	6.66×10^{-15}	9.33×10^{-16}
GO:0016192 vesicle-mediated transport	93	2.03×10^{-13}	2.48×10^{-14}
GO:0006629 lipid metabolic process	155	2.15×10^{-11}	2.17×10^{-12}
GO:0008152 metabolic process	156	2.22×10^{-11}	2.17×10^{-12}

Table 4.5 Top significant GO-CC term enriched (FDR sorted)

GO-CC Term Name	#genes	Enriched Adjusted p-value	FDR
GO:0070062 extracellular exosome	1045	1.09×10^{-171}	9.37×10^{-172}
GO:0005739 mitochondrion	632	6.43×10^{-82}	2.76×10^{-82}
GO:0030529 intracellular ribonucleoprotein complex	209	7.92×10^{-76}	2.26×10^{-76}
GO:0005737 cytoplasm	1669	1.38×10^{-74}	2.96×10^{-75}
GO:0005829 cytosol	609	2.06×10^{-64}	3.53×10^{-65}
GO:0005840 ribosome	131	2.87×10^{-51}	4.10×10^{-52}
GO:0016020 membrane	1646	9.62×10^{-49}	1.18×10^{-49}
GO:0005634 nucleus	1453	2.65×10^{-47}	2.84×10^{-48}
GO:0005654 nucleoplasm	590	2.30×10^{-43}	2.19×10^{-44}
GO:0005743 mitochondrial inner membrane	192	2.92×10^{-43}	2.50×10^{-44}

*See supplementary Table S5 for details

Table 4.6 Top significant GO-MF term enriched (FDR sorted)

GO-MF Term Name	#genes	Enriched Adjusted p-value	FDR
GO:0044822 poly(A) RNA binding	538	2.94×10^{-120}	2.82×10^{-120}
GO:0098641 cadherin binding involved in cell-cell adhesion	158	6.77×10^{-43}	3.25×10^{-43}
GO:0005515 protein binding	1046	1.15×10^{-36}	3.68×10^{-37}
GO:0003735 structural constituent of ribosome	137	2.83×10^{-31}	6.29×10^{-32}
GO:0003723 RNA binding	285	3.28×10^{-31}	6.29×10^{-32}
GO:0000166 nucleotide binding	551	8.93×10^{-28}	1.43×10^{-28}
GO:0019899 enzyme binding	143	3.90×10^{-15}	5.34×10^{-16}
GO:0016491 oxidoreductase activity	196	5.39×10^{-14}	6.46×10^{-15}
GO:0032403 protein complex binding	131	7.38×10^{-13}	8.61×10^{-14}
GO:0019904 protein domain specific binding	111	4.18×10^{-12}	3.95×10^{-13}

of 1.02×10^{-39}, was one of the most important GO-BP terms. This GO-BP keyword was linked to 195 genes with an enriched corrected p-value of 1.04×10^{-39}. The remaining terms are included in Table 4.4 . We included a supplementary file with a complete list of all GO-BP terms. As one of the most significant GO-CC terms, *GO:0070062 extracellular exosome* had an FDR value of 9.37×10^{-172}. 1045 genes in total had enriched adjusted p-values of 1.09×10^{-171} connected with this GO-CC term. Table 4.5

displays the remaining terms. The whole set of GO-CC terms is available in a supplemental file. Last but not least, *GO:0044822 poly(A) RNA binding*, with a minimum FDR value of 2.82×10^{-120}, was one of the most significant GO-MF terms. With an enriched corrected p-value of 2.94×10^{-120}, a total of 538 genes were linked to this GO-MF term. Table 4.6 provides information. As a supplementary file, we gave the whole set of GO-MF terms.

4.4 Conclusion and Future Work

In this chapter, we put forward a method for discovering cluster-specific frequent biomarkers in single-cell sequencing data utilising dimensional reduction and cell clustering. In order to create the framework, we first filtered our single-cell RNA sequencing dataset by removing features and cells with small numbers of cells and features. Then, we saved the data and the findings of analysis as a Seurat object and carried out a number of analysis steps, including filtering cells based on quality control metrics, removing poor-quality or dying cells, computing cell-to-cell highly variable features from the dataset, and using PCA to project high-dimensional data into an ideal low-dimensional space. Based on the substantial enrichment of low p-value features, we found fifty "significant" principal components (PCs). We then applied the Louvain agglomerative graph-based clustering technique on the cell of the first 10 PCs and obtained 10 clusters with a maximum modularity of 0.885. Then, using downstream analysis and the statistical test "MAST", we discovered 3871 biomarkers specific to clusters by only taking up-regulated differentially expressed genes (DEGs) into consideration as cluster markers, with a log_2FC threshold of 0.25 and a minimum percentage of feature detection of 25%. We discovered 1892 most frequent markers, or overlapping biomarkers, from these cluster-specified biomarkers.

After that, the five highest degree genes ($Rps4x$, $Rps18$, $Rpl13a$, $Rps12$ and $Rpl18a$) were identified using degree hub gene network analysis using *Cytoscape* online tool (version 3.8.0). It is interesting to see that the most of our top hub genes are ribosomal protein genes. Ribosomal protein genes are the most highly expressed genes in the majority of cell types, and they provide the biological justification for the top hub genes in single cell analysis. Genes encoding ribosomal proteins are essential for cell proliferation and expansion [33].

In addition, we employed pathway analysis through David 6.8 software on markers associated with specific clusters. To sum up, our framework is helpful for biological interpretation of single-cell sequencing data analysis and effectively detecting the overlapping biomarkers specific to clusters. We may highlight one benefit of our work, which is the recent emergence of certain novel methodologies in the evolving field of single cell sequencing analysis. Each technology has unique strengths and limitations, and measurements are only based on specific cellular identity traits, which drives the need to aggregate data from several datasets to improve the interpretation of other datasets. For instance, single cell ATAC-seq (scATAC-seq) can disclose enhancer areas and regulatory logic in a unique way, but it is presently not able to attain the same capacity for unsupervised cell type discovery as with transcriptomics [34]. The STARmap method, on the other hand, allows for the measurement of more than 1,000 genes in spatially intact tissue, but it predicts that this number of genes will serve as the maximum limit for such methods without super-resolution microscopy or the physical expansion of hydrogels [35]. In our framework, we have made an effort to create efficient single cell dataset tools that can support similarly massive improvements in single cell data analysis and interpretation.

Our methodology has a flaw in that we are using raw data matrix here instead of an imputed matrix. We may use an imputed data matrix to enhance our framework in the upcoming study. In order to improve the clustering method, we may also expand on our present work by incorporating multi-objective optimization techniques. It can be used for big data analysis and the discovery of rare cells in single-cell RNA sequencing data.

References

[1] Tang, X., Huang, Y., Lei, J. et al., The Single-cell Sequencing: New Developments and Medical Applications. Cell Biosci 9, 53 (2019).
[2] Ruth Huh, Yuchen Yang, Yuchao Jiang, Yin Shen, Yun Li, SAME-clustering: Single-cell Aggregated Clustering via Mixture Model Ensemble, Nucleic Acids Research, Volume 48, Issue 1, 10 January 2020, Pages 86-95.
[3] Feng C, Liu S, Zhang H, et al, Dimension Reduction and Clustering Models for Single-Cell RNA Sequencing Data: A Comparative Study, Int J Mol Sci. 2020;21(6):2181, Published 2020 Mar 22.
[4] Qiu, X., Hill, A., Packer, J., Lin, D., Ma, Y.-A., Trapnell, C., Single-cell mRNA Quantification and Differential Analysis with Census, Nat. Methods 2017, 14, 309-315.

[5] Wu, X., Kumar, V., Ross Quinlan, J., Ghosh, J., Yang, Q., Motoda, H., McLachlan, G.J., Ng, A., Liu, B., Yu, P.S., et al., Top 10 Algorithms in Data-mining. Knowl. Inf. Syst. 2008, 14, 1-37.

[6] Zeisel, A., Munoz-Manchado, A.B., Codeluppi, S., Lönnerberg, P., Manno, G.L., Jureus, A., Marques, S., Munguba, H., He, L., Betsholtz, C., et al., Cell Types in The Mouse Cortex and Hippocampus Revealed by Single-cell RNA-seq., Science 2015, 347, 1138-1142.

[7] Zurauskien · e, J., Yau, C., pcaReduce: Hierarchical Clustering of Single-cell Transcriptional Profiles., BMC Bioinform. 2016, 17, 140.

[8] Bartenhagen, C., Klein, HU., Ruckert, C. et al., Comparative Study of Unsupervised Dimension Reduction Techniques for The Visualization of Microarray Gene Expression Data., BMC Bioinformatics 11, 567 (2010).

[9] Grün, D., Lyubimova, A., Kester, L. et al., Single-cell Messenger RNA Sequencing Reveals Rare Intestinal Cell Types, Nature 525, 251-255 (2015).

[10] Butler A, Hoffman P, Smibert P, Papalexi E, Satija R (2018), Integrating Single-cell Transcriptomic Data Across Different Conditions, Technologies, and Species, Nature Biotechnology, 36, 411-420.

[11] Blondel, V., Guillaume, J., Lambiotte, R., Lefebvre, E. (2008)., Fast Unfolding of Communities in Large Networks, Journal of Statistical Mechanics: Theory and Experiment, 2008, 10008.

[12] Finak, G., McDavid, A., Yajima, M. et al., MAST: A Flexible Statistical Framework for Assessing Transcriptional Changes and Characterizing Heterogeneity in Single-cell RNA Sequencing Data, Genome Biol 16, 278 (2015).

[13] Dal Molin A, Baruzzo G, Di Camillo B., Single-Cell RNA-Sequencing: Assessment of Differential Expression Analysis Methods, Front Genet. 2017;8:62, Published 2017 May 23.

[14] Shannon, P.; Markiel, A.; Ozier, O.; Baliga, N.S.; Wang, J.T.; Ramage, D.; Amin, N.; Schwikowski, B.; Ideker, T. Cytoscape: A Software Environment for Integrated Models of Biomolecular Interaction Networks. *Genome Res.* **2003**, *13*, 2498-2504.

[15] Dennis, G., Sherman, B.T., Hosack, D.A. et al., DAVID: Database for Annotation, Visualization, and Integrated Discovery, Genome Biol 4, R60 (2003).

[16] Stuart T, Butler A, Hoffman P, Hafemeister C, Papalexi E, III WMM, Hao Y, Stoeckius M, Smibert P, Satija R (2019), Comprehensive Integration of Single-Cell Data, Cell, 177, 1888-1902.

[17] Ilicic T, Kim JK, Kolodziejczyk AA, et al., Classification of Low Quality Cells From Single-cell RNA-seq Data, Genome Biol., 2016;17:29, Published 2016 Feb 17.

[18] Brennecke, P., Anders, S., Kim, J. et al., Accounting for Technical Noise in Single-cell RNA-seq Experiments, Nat Methods 10, 1093–1095 (2013).

[19] Stuart T, Butler A, Hoffman P, Hafemeister C, Papalexi E, III WMM, Hao Y, Stoeckius M, Smibert P, Satija R, Comprehensive Integration of Single-cell Data, bioRxiv 460147.

[20] Cunningham, J. and Zoubin Ghahramani, Linear Dimensionality Reduction: Survey, Insights, and Generalizations, J. Mach. Learn. Res. 16 (2015): 2859-2900.

[21] Jolliffe Ian T. and Cadima Jorge, 2016, Principal Component Analysis: A Review and Recent Developments, Phil. Trans. R. Soc. A.3742015020220150202.

[22] Macosko EZ, Basu A, Satija R, Nemesh J, Shekhar K, Goldman M, Tirosh I, Bialas AR, Kamitaki N, Martersteck EM, Trombetta JJ, Weitz DA, Sanes JR, Shalek AK, Regev A, McCarroll SA, Highly Parallel Genome-wide Expression Profiling of Individual Cells Using Nanoliter Droplets, Cell. 2015 May 21;161(5):1202-1214, PMID: 26000488; PMCID: PMC4481139.

[23] M. E. J. Newman, Modularity and Community Structure in Networks, Proceedings of the National Academy of Sciences, Jun 2006, 103 (23) 8577-8582.

[24] Kirianovskii I., Granichin O., Proskurnikov A., A New Randomized Algorithm for Community Detection in Large Networks, ELSEVIRE, Volume 49, Issue 13, 2016, Pages 31-35, ISSN 2405-8963.

[25] E. Myasnikov, Using UMAP for Dimensionality Reduction of Hyperspectral Data, 2020 International Multi-Conference on Industrial Engineering and Modern Technologies (FarEastCon), 2020, pp. 1-5.

[26] An, X., Schulz, V. P., Li, J., Wu, K., Liu, J., Xue, F., Hu, J., Mohandas, N., & Gallagher, P. G. (2014)., Global Transcriptome Analyses of Human and Murine Terminal Erythroid Differentiation, Blood, 123(22), 3466-3477.

[27] Bradley P Balaton, Thomas Dixon-McDougall, Samantha B Peeters, Carolyn J Brown, The eXceptional Nature of The X Chromosome, Human Molecular Genetics, Volume 27, Issue R2, 01 August 2018, Pages R242-R249.

[28] Mar Matarin, Dervis A. Salih, Marina Yasvoina, Damian M. Cummings, Sebastian Guelfi, Wenfei Liu, Muzammil A., Nahaboo Solim, Thomas G. Moens, Rocio Moreno Paublete, Shabinah S. Ali, Marina Perona, Roshni Desai, Kenneth J. Smith, Judy Latcham, Michael Fulleylove, Jill C. Richardson, John Hardy, Frances A. Edwards, A Genome-wide Gene-Expression Analysis and Database in Transgenic Mice during Development of Amyloid or Tau Pathology, A Cell Press journal, VOLUME 10, ISSUE 4, P633-644, FEBRUARY 03, 2015.

[29] Wright Muelas, M., Mughal, F., O′Hagan, S. et al., The Role and Robustness of The Gini Coefficient As An Unbiased Tool for The Selection of Gini Genes for Normalising Expression Profiling Data, Sci Rep 9, 17960 (2019).

[30] Fa, B., Wei, T., Zhou, Y. et al., GapClust is a Light-weight Approach Distinguishing Rare Cells from Voluminous Single-cell Expression Profiles, Nat Commun 12, 4197 (2021).

[31] Wisdom, A.J., Mowery, Y.M., Hong, C.S. et al., Single-cell Analysis Reveals Distinct Immune Landscapes in Transplant and Primary Sarcomas That Determine Response or Resistance to Immunotherapy, Nat Commun 11, 6410 (2020).

[32] Onur Basak, Teresa G. Krieger, Mauro J. Muraro, Kay Wiebrands, Daniel E. Stange, Javier Frias-Aldeguer, Nicolas C. Rivron, Marc van de Wetering, Johan H. van Es, Alexander van Oudenaarden, Benjamin D. Simons, and Hans Clevers, Troy+ Brain Stem Cells Cycle Through Quiescence and Regulate Their Number by Sensing Niche Occupancy, PNAS January 23, 2018 115 (4) E610-E619; first published January 8, 2018.

[33] Petibon C, Malik Ghulam M, Catala M, Abou Elela S. Regulation of ribosomal protein genes: An ordered anarchy. Wiley Interdiscip Rev RNA. 2021;12(3):e1632.

[34] Lake, B. B. et al. Integrative single-cell analysis of transcriptional and epigenetic states in the human adult brain. Nat. Biotechnol. 36, 70-80.

[35] Wang, X. et al. Three-dimensional intact-tissue sequencing of single-cell transcriptional states. Science 361 (2018).

[36] Seth S, Mallik S, Bhadra T, Zhao Z, Dimensionality Reduction and Louvain Agglomerative Hierarchical Clustering for Cluster-Specified Frequent Biomarker Discovery in Single-Cell Sequencing Data, Front Genet. 2022 Feb 7;13:828479, doi: 10.3389/fgene.2022.828479, PMID: 35198011; PMCID: PMC8859265.

5

Application of Detecting Discriminant Features from Stationary Nucleotide Base Pattern to the Classification of Essential Genes

5.1 Introduction

Essential genes are regarded as the foundational elements of life because they are required for an organism to survive. The key to a comprehensive understanding of life and extension may lie in essential genes, which contain the essential information found in genomes [1]. It determines significant features related to cellular structure, chemistry, and reproduction, among other things. Genomes have stored instructions for the common biological processes that are found in all living things, and the data may be species-specific. Some genes seem to be necessary for survival, while others appear to be optional. To extract genes and define the basic components necessary to sustain cellular life, essential genes have been provided. A cell would expire if any of these genes were absent. Consequently, essential gene prediction helps in determining the absolute minimum of genes required for the survival of specific cell types. Our understanding of the origins of life is aided by discovery and analysis of significant genes [2]. Additionally, essential genes are critical to synthetic molecular biology, which is crucial to genome development. A thorough understanding of key genes can enable researchers to clarify the biological nature of microorganisms [5], produce the tiniest genome subset [6], develop interesting therapeutic targets, and develop potential cures for infectious diseases [7]. The identification of essential genes has been considered crucial in bioinformatics and genomics due to their significance.

An ensemble of essential genes are required for an organism to survive and reproduce in a specific environment. Most of these are only required in certain situations. As an example, the gene that produces lysine is not required if the amino acid is supplied to the organism. The gene encoding the enzyme responsible for lysine biosynthesis, however, becomes crucial in the absence of an amino acid supply since protein synthesis is impossible without it. In almost each species, essential genes control the functioning of fundamental cells [8, 9]. In situations where the majority of the required nutrients are present in the growth medium and the organism grows at its optimum temperature, genes are considered essential if they cannot be knocked out independently. Identifying the essential genes among the observed genes is one of the primary challenges. Several experimental methods, including gene knockouts [10], RNA interference [11], transposon mutagenesis [12], and single-gene knockout procedures [13], are available to detect essential genes in microorganisms. However, these experimental methods offer a number of advantages and are generally beneficial. Still they are costly and time-consuming. Therefore, computational techniques are desired to detect essential genes.

Numerous computer techniques, in particular machine learning techniques, have been used to identify essential genes because of their biological significance. Several feature extraction and model building algorithms have been developed for achieving this objective [17, 18, 19, 20]. The authors in [21] successfully assessed protein dispensability in *Saccharomyces cerevisiae* using high throughput data and machine learning approaches. Using a number of intrinsic genetic features, the authors in [22] built a machine learning model to predict essential genes in *Saccharomyces cerevisiae*. Furthermore, to identify knockdown lethality in mice, the authors of the existing study [23] developed three machine learning methods based on instructive genetic traits. An essential gene classification approach was presented by the authors in [24] using hybrid traits including intrinsic and context-dependent genomic features. When predicting cross organisms using ten-fold cross-validation, this model achieved area under the receiver operating characteristic curve (AUC) scores of 0.69–0.89 and 0.86–0.93 when testing the same organism.

The authors in [25] made major contributions by combining sequence- and network-based features to identify significant genes and by using a deep-learning-based model to understand the characteristics derived from sequencing data and protein–protein interaction networks to arrive at reliable conclusions. Using cross-validation, paired, self-test, and leave-one-species-out experiments, authors of [26] reported the results of extensive research on 31 bacterial species. Purine and pyrimidine distribution, among other quantitative techniques, were used by the authors in [27] to propose an approach for identifying the crucial genes of four species. A paradigm for finding essential genes using an ensemble deep neural network was put out in this previous study [4]. In the article [28], the authors created an artificial neural network method for predicting prokaryotic essential genes based on sequence-based features. For the thorough investigation of human essential genes, authors in [29] created the Human Essential Genes Interactive investigation Platform (HEGIAP), a web server.

To identify the essential human genes, authors in [30] created an expression-based predictor. In order to predict lncRNAs in cancer cells, the predictor used gene expression profiles. This existing study [31] developed the essential gene database NetGenes, which includes predictions for 2711 bacterial species based on network-based features. The STRING database was analyzed for features using the protein–protein interaction network. In *Tribolium castaneum* and *Drosophila melanogaster*, researchers [32] estimated the essential genes based on physicochemical and statistical data as well as subcellular locations. From the essential and non-essential data, they extracted extrinsic and intrinsic traits. In order to find essential genes, this study assessed the DNA sequences of five different species, including *Homo sapiens, Danio rerio, Drosophila melanogaster, Mus musculus, and Arabidopsis thaliana*. In order to discover any informative patterns that differentiate the species, the proposed method extracts co-occurrence matrices from the sequences of the essential genes [45]. The impact of various co-occurrence matrices and existing features, such as Hurst exponent, fractal dimension, Shannon entropy, and modified Shannon entropy, is also explored in this chapter.

5.2 Basic Terminology

A family of genes known as essential genes is associated with key cellular functions necessary for species survival. The multidisciplinary process of identifying essential genes requires both computational and wet-lab validation investigations. To increase classification accuracy, a number of machine learning techniques have been developed, making the process time- and resource-intensive. As a result, most of these methods involve supervised learning techniques, which require a significant amount of labeled training data and are therefore generally troubling for less sequenced species. On the other hand, there is an abundance of unlabeled essential gene sequence data as a result of the proliferation of high-throughput wet-lab experimental techniques like next-generation sequencing. In the initial research, it was found that each DNA sequence is represented by a fixed-dimensional feature vector employing a variety of quantitative metrics, including Hurst exponent, Shannon entropy, modified Shannon entropy, and fractal dimension. We transform gene sequences into binary sequences based on the distribution of pyrimidine and purine in order to estimate these quantitative measures. Nitrogenous bases make up the two main types of nucleotide bases in DNA. While cytosine (C) and thymine (T) are pyrimidines, adenine (A) and guanine (G) are purines. Here, the bases of purine and pyrimidine are denoted by 1 and 0, respectively.

$$A/G \longrightarrow 1 \quad \text{and} \quad C/T \longrightarrow 0. \tag{5.1}$$

5.2.1 Shannon entropy and modified Shannon entropy

To estimate how much uncertainty or information a sequence holds, one can apply Shannon entropy (SE) [33, 14]. The distribution of each word is impacted by the uncertainty. Base pair uncertainty for a sequence ranges from 0 to $2n$ where n is the length of a word. In order to determine information entropy, the SE employs the probability p of the two possibilities (0/1). The SE of a binary sequence is given by the following equation:

$$\text{SE} = -\sum_{i=0}^{1} p_i \log_2(p_i). \tag{5.2}$$

The SE is used to calculate the uncertainty in a binary string, where p_i is the probability of two values in a binary sequence [16]. The event is guaranteed to never occur when the probability is $p = 0$, producing no uncertainty and an entropy of 0. Similarly, if $p = 1$, the outcome is definite and the entropy must be zero. The uncertainty is maximum and the SE is 1 when $p = 1/2$.

With respect to various word sizes, the modified Shannon entropy (MSE) is provided by

$$\text{MSE} = -\sum_{j=1}^{k} w_j \log_2(w_j), \tag{5.3}$$

where w_j denotes how often the *j*th word appears in the gene sequence. For a word of length 1, the frequency of purine or pyrimidine 0, 1, and for a word of length 2, the two-time repeat of purine or pyrimidine 00, 10, 01, 11 are used to calculate w_j, respectively. k is the number of words resulting from combining the maximum lengths of the purine and pyrimidine bases [27].

5.2.2 Hurst exponent

The smoothness and degree of similarity of a dataset are assessed using the Hurst exponent (HE). When analyzing auto-correlation in time series analysis, the HE is often applied. It has a value between 0 and 1 and is determined using rescaled range analysis (R/S analysis) [34, 15]. An HE value between 0 and 0.5 denotes a negative autocorrelation of a time series, while an HE between 0.5 and 1 denotes a positive autocorrelation. The series is random if the HE value is 0.5, which means that there is no correlation between the variable and its prior values [35, 36]. The following equation can be used to calculate the HE of a binary sequence D_n:

$$\frac{R(n)}{S(n)} = (\frac{n}{2})^{\mathrm{HE}}, \tag{5.4}$$

where

$$S(n) = \sqrt{\frac{1}{n}\sum_{i=1}^{n}(D_i - m)^2} \tag{5.5}$$

and

$$R(n) = \max(X_1, X_2, ..., X_n) - \min(X_1, X_2, ..., X_n) \tag{5.6}$$

$$X_t = \sum_{i=1}^{t}(D_i - m) \quad \text{for } t = 1, 2, 3, ..., n \tag{5.7}$$

$$m = \frac{1}{n}\sum_{i=1}^{n} D_i. \tag{5.8}$$

5.2.3 Fractal dimension

Indicator matrices are created from every DNA sequence [37, 38]. Let *D(N)* be a DNA sequence with four symbols from *X* of length *N* and *X* = *{A, T, C, G}* denoting the set of finite alphabet nucleotides. The following equation describes the indicator function for each DNA sequence:

$$F : D(N) \times D(N) \rightarrow \{0, 1\}, \text{and } D(N) = \{0, 1\} \tag{5.9}$$

such that the indicator matrix will be

$$I(N, N) = \begin{cases} 1, & \text{if } s_i = s_j \\ 0, & \text{if } s_i \neq s_j \end{cases} \quad \text{where } s_i, s_j \in D(N). \tag{5.10}$$

In this case, a binary image of the DNA sequence is produced as a 2D dot-plot by the matrix *I(N, N)*, which has values of 0 and 1. The binary image can depict the distribution of zeros and ones within the same sequence. You may designate a white dot as 0, and a black dot as 1. The average number of $\sigma(n)$ of 1, randomly selected $n \times n$ from a $N \times N$ indicator matrix, can be used to calculate the fractal dimension (FD) from an indicator matrix [39, 40, 41]. The following equation computes the FD using $\sigma(n)$:

$$\mathrm{FD} = -\frac{1}{N}\sum_{n=2}^{N}\frac{\log(\sigma(n))}{\log n}. \tag{5.11}$$

5.3 Proposed Scheme

For the sake of experimental results and discussion of this chapter, we used the Database of Essential Genes (http://www.essentialgene.org/). Five different species' necessary genes are included in this dataset. There are 2051 essential genes for *Homo sapiens* (HS), 315 for *Danio rerio* (DR), 339 for *Drosophila melanogaster* (DOM), 356 for *Arabidopsis thaliana* (AT), and 125 for *Mus musculus* (MM). For reference, Table 5.1 provides a list of some of the terms used in the proposed approach [45].

Table 5.1 List of species considered in the proposed technique.

Name	Symbol used
Arabidopsis thaliana	AT
Drosophila melanogaster	DOM
Danio rerio	DR
Homo sapiens	HS
Mus musculus	MM
Naming convention for *Arabidopsis thaliana*	[AT_1-AT_{356}]
Naming convention for *Drosophila melanogaster*	[DOM_1-DOM_{339}]
Naming convention for *Danio rerio*	[DR_1-DR_{315}]
Naming convention for *Homo sapiens*	[HS_1-HS_{2051}]
Naming convention for *Mus musculus*	[MM_1-MM_{125}]

5.3.1 Proposed feature representation technique

The four nucleotides adenine (A), guanine (G), cytosine (C), and thymine (T) form the DNA (deoxyribonucleic acid) sequence of the essential gene S. As a result, there might be numerous instances of the sequence S that contain combinations of A, C, T, G. The relationship between the nucleotides is brought out by the co-occurrences of A, C, T, G in the DNA sequence. The process for determining the co-occurrences of the nucleotides A, C, T, G within S has been proposed for the first time. Finding these co-occurrences serves the purpose of analyzing the A, C, T, G patterns within DNA sequence S in order to extract certain useful traits that serve to distinguish the species by their essential genes in a distinctive way. Let's assume x represents a vector of nucleotides, consisting of (A, C, T, G). The co-occurrence matrices created by the vector combination displayed in Table 5.2 illustrate the possibility of incorporating these characters in the DNA gene sequences.

Table 5.2 Possible sets of occurrences of nucleobases A, C, T, G in a DNA sequence or essential gene formed by the combination of vectors, where I, J, K, L, M, N, O, P are the co-occurrence matrices.

X	Y	$X^T \times Y$
$X_1 = (A, C, T, G)$	(A, C, T, G)	$\underset{4\times4}{I} = \underset{4\times1}{X_1{}^T} \times \underset{1\times4}{Y}$
$X_2 = (AA, CC, TT, GG)$	(A, C, T, G)	$\underset{4\times4}{J} = \underset{4\times1}{X_2{}^T} \times \underset{1\times4}{Y}$
$X_3 = (AC, AT, AG, CT, CG, TG)$	(A, C, T, G)	$\underset{6\times4}{K} = \underset{6\times1}{X_3{}^T} \times \underset{1\times4}{Y}$
$X_4 = (CA, TA, GA, TC, GC, GT)$	(A, C, T, G)	$\underset{6\times4}{L} = \underset{4\times1}{X_4{}^T} \times \underset{1\times4}{Y}$
$X_5 = (ACT, ACG, ATG, CTG)$	(A, C, T, G)	$\underset{4\times4}{M} = \underset{4\times1}{X_5{}^T} \times \underset{1\times4}{Y}$
$X_6 = (CAT, CAG, TAG, TCG)$	(A, C, T, G)	$\underset{4\times4}{N} = \underset{4\times1}{X_6{}^T} \times \underset{1\times4}{Y}$
$X_7 = (ATC, AGC, AGT, CGT)$	(A, C, T, G)	$\underset{4\times4}{O} = \underset{4\times1}{X_7{}^T} \times \underset{1\times4}{Y}$
$X_8 = (TCA, GCA, GTA, GTC)$	(A, C, T, G)	$\underset{4\times4}{P} = \underset{4\times1}{X_8{}^T} \times \underset{1\times4}{Y}$

Here, the computed co-occurrence matrices of various nucleobase combinations show how the nucleobases are distributed across the critical gene S. This nucleobase distribution analyzes the textural pattern and takes into account the spatial relationships between the nucleobases in the essential gene S. Experimental findings show that the spatial relationships between nucleobases cannot reliably reveal information on both stationary and non-stationary patterns of A, C, T, and G. But the spatial link that was discovered includes both of these details simultaneously. Therefore, rather than distinguishing stationary and non-stationary patterns in S, it is simpler statistically to compute information while taking both stationary and non-stationary patterns into account at the same time. Any organism must have its essential genes in order to survive. It promotes cell development. The length of each gene sequence varies, and the A, C, T, G nucleobase configurations are zigzagged. Therefore, identifying the stationary and non-stationary patterns of A, C, T, G as well as the co-occurrences of the various combinations of these nucleobases would aid in determining the natural pattern of genes. In order to properly analyze the statistical and structural information from those patterns, it is necessary to derive the valuable patterns of the variety of A, C, T, G using co-occurrence matrix descriptors. We used the concepts of a gray-level co-occurrence matrix as a result of being inspired by the co-occurrence matrix of texture analysis [3] in image processing and pattern recognition. From each essential gene data, we have calculated a number of co-occurrence matrices here. Now $I_{4\times4}$, $J_{4\times4}$, $K_{6\times4}$, $L_{6\times4}$, $M_{4\times4}$, $N_{4\times4}$, $O_{4\times4}$, and $P_{4\times4}$ co-occurrence matrices with various patterns of A, C, T, G nucleobases in each DNA sequence are generated. In Tables 5.3, 5.4, 5.5, 5.6, 5.7, 5.8, 5.9, and 5.10, certain co-occurrence matrices are defined respectively.

Table 5.3 Co-occurrence matrix I that contain several patterns of A, C, T, G nucleobases in DNA gene sequence S.

	A	C	T	G
A	#(AA)	#(AC)	#(AT)	#(AG)
C	#(CA)	#(CC)	#(CT)	#(CG)
T	#(TA)	#(TC)	#(TT)	#(TG)
G	#(GA)	#(GC)	#(GT)	#(GG)

Table 5.4 Co-occurrence matrix J that contains several patterns of A, C, T, G nucleobases in DNA gene sequence S.

	A	C	T	G
AA	#(AAA)	#(AAC)	#(AAT)	#(AAG)
CC	#(CCA)	#(CCC)	#(CCT)	#(CCG)
TT	#(TTA)	#(TTC)	#(TTT)	#(TTG)
GG	#(GGA)	#(GGC)	#(GGT)	#(GGG)

Table 5.5 Co-occurrence matrix K that contains several patterns of A, C, T, G nucleobases in DNA gene sequence S.

	A	C	T	G
AC	#(ACA)	#(ACC)	#(ACT)	#(ACG)
AT	#(ATA)	#(ATC)	#(ATT)	#(ATG)
AG	#(AGA)	#(AGC)	#(AGT)	#(AGG)
CT	#(CTA)	#(CTC)	#(CTT)	#(CTG)
CG	#(CGA)	#(CGC)	#(CGT)	#(CGG)
TG	#(TGA)	#(TGC)	#(TGT)	#(TGG)

Table 5.6 Co-occurrence matrix L that contains several patterns of A, C, T, G nucleobases in DNA gene sequence $\mathcal{S}$.

	A	C	T	G
CA	#(CAA)	#(CAC)	#(CAT)	#(CAG)
TA	#(TAA)	#(TAC)	#(TAT)	#(TAG)
GA	#(GAA)	#(GAC)	#(GAT)	#(GAG)
TC	#(TCA)	#(TCC)	#(TCT)	#(TCG)
GC	#(GCA)	#(GCC)	#(GCT)	#(GCG)
GT	#(GTA)	#(GTC)	#(GTT)	#(GTG)

Table 5.7 Co-occurrence matrix M that contains several patterns of A, C, T, G nucleobases in DNA gene sequence $\mathcal{S}$/

	A	C	T	G
ACT	#(ACTA)	#(ACTC)	#(ACTT)	#(ACTG)
ACG	#(ACGA)	#(ACGC)	#(ACGT)	#(ACGG)
ATG	#(ATGA)	#(ATGC)	#(ATGT)	#(ATGG)
CTG	#(CTGA)	#(CTGC)	#(CTGT)	#(CTGG)

Table 5.8 Co-occurrence matrix N that contains several patterns of A, C, T, G nucleobases in DNA gene sequence $\mathcal{S}$.

	A	C	T	G
CAT	#(CATA)	#(CATC)	#(CATT)	#(CATG)
CAG	#(CAGA)	#(CAGC)	#(CAGT)	#(CAGG)
TAG	#(TAGA)	#(TAGC)	#(TAGT)	#(TAGG)
TCG	#(TCGA)	#(TCGC)	#(TCGT)	#(TCGG)

Table 5.9 Co-occurrence matrix O that contains several patterns of A, C, T, G nucleobases in DNA gene sequence $\mathcal{S}$.

	A	C	T	G
ATC	#(ATCA)	#(ATCC)	#(ATCT)	#(ATCG)
AGC	#(AGCA)	#(AGCC)	#(AGCT)	#(AGCG)
AGT	#(AGTA)	#(AGTC)	#(AGTT)	#(AGTG)
CGT	#(CGTA)	#(CGTC)	#(CGTT)	#(CGTG)

Table 5.10 Co-occurrence matrix P that contains several patterns of A, C, T, G nucleobases in DNA gene sequence $\mathcal{S}$.

	A	C	T	G
TCA	#(TCAA)	#(TCAC)	#(TCAT)	#(TCAG)
GCA	#(GCAA)	#(GCAC)	#(GCAT)	#(GCAG)
GTA	#(GTAA)	#(GTAC)	#(GTAT)	#(GTAG)
GTC	#(GTCA)	#(GTCC)	#(GTCT)	#(GTCG)

The aforementioned co-occurrence matrices are produced here from the given DNA sequence $\mathcal{S}$. The number of times that each of the nucleobases A, C, T, G have been found in a given combination and offset in $\mathcal{S}$ is contained in the co-occurrence matrix $\mathcal{G}$. A co-occurrence matrix of size $q \times 4$ is produced for the specified offset by a sequence $\mathcal{S}$ that contains q different combinations of the A, C, T, G nucleobases. Therefore, the (r, s)th value of a co-occurrence matrix (Tables 5.3–5.10) indicates how frequently rth and sth nucleobases are present in $\mathcal{S}$. Therefore, each co-occurrence matrix (Tables 5.3–5.10)in this instancc is mathematically provided by

$$\mathcal{G} = \sum_{i=1}^{n}\sum_{j=1}^{n} \begin{cases} 1 & G_{(i,j)} = r \ \& \ G_{(i+\triangle i, j+\triangle j)} = s \\ 0 & \text{otherwise} \end{cases} . \tag{5.12}$$

The spatial relation for which the matrix $\mathcal{G}$ is calculated is defined by the offset $(\triangle i, \triangle j)$. The number of A, C, T, G combinations that co-occur in $\mathcal{S}$ is determined by co-occurrence matrices. Each matrix $\mathcal{G}$ is therefore normalized to $\mathcal{G}' = \frac{\mathcal{G}}{\sum_{r=0}^{q}\sum_{s=0}^{q}\mathcal{G}(r,s)}$ in order to extract distinguishing and discriminating features. Then, certain features, including entropy, dissimilarity, energy, homogeneity, and contrast, are computed using the normalized co-occurrence matrix $\mathcal{G}'$. The definitions of these features in mathematics are displayed in Table 5.11.

Table 5.11 Features extracted from a co-occurrence matrix $\mathcal{G}$ of DNA sequence S.

Feature	Formulae
Energy	$\sum_{r=0}^{q}\sum_{s=0}^{q}\mathcal{G}'(r,s)^2$
Entropy	$\sum_{r=0}^{q}\sum_{s=0}^{q} -\mathcal{G}'(r,s) \times \ln(\mathcal{G}'(r,s))$
Homogeneity	$\sum_{r=0}^{q}\sum_{s=0}^{q}\frac{\mathcal{G}'(r,s)}{(1+(r-s)^2)}$
Contrast	$\sum_{r=0}^{q}\sum_{s=0}^{q}\mathcal{G}'(r,s) \times (r-s)^2$
Dissimilarity	$\sum_{r=0}^{q}\sum_{s=0}^{q}\mathcal{G}'(r,s) \times \lvert(r-s)\rvert$

After the extraction of the features defined in Table 5.11, from each co-occurrence matrix (Tables 5.3–5.10), the list of feature vectors generated from these matrices is as follows:

$f_I = (f_1, f_2, f_3, f_4, f_5)$ from I (Table 5.3)
$f_J = (f_6, f_7, f_8, f_9, f_{10})$ from J (Table 5.4)
$f_K = (f_{11}, f_{12}, f_{13}, f_{14}, f_{15})$ from K (Table 5.5)

$f_L = (f_{16}, f_{17}, f_{18}, f_{19}, f_{20})$ from L (Table 5.6)
$f_M = (f_{21}, f_{22}, f_{23}, f_{24}, f_{25})$ from M (Table 5.7)
$f_N = (f_{26}, f_{27}, f_{28}, f_{29}, f_{30})$ from N (Table 5.8)
$f_O = (f_{31}, f_{32}, f_{33}, f_{34}, f_{35})$ from O (Table 5.9)
$f_P = (f_{36}, f_{37}, f_{38}, f_{39}, f_{40})$ from P (Table 5.10)

Consequently, the feature vector $f = (f_I, f_J, f_K, f_L, f_M, f_N, f_O, f_P)$ provides the final feature representation of a DNA sequence or essential gene $\mathcal{S}$.

5.3.2 Classification

In this study, the decision tree (DT), k-nearest neighbor (KNN), and support vector machine (SVM) classifiers are used for the classification of the essential genes in the employed species. The dataset of each species, including those of *Arabidopsis thaliana* (AT), *Drosophila melanogaster* (DOM), *Danio rerio* (DR), *Homo sapiens* (HS), and *Mus musculus* (MM), is equally split into a training set and a testing set during the experiment. After that, a five-fold cross-validation method is used. The average performance for the testing data for the proposed framework is subsequently addressed.

Decision tree: The Iterative Dichotomiser 3 algorithm (ID3) or the CART method *(Classification and Regression Tree)* [42] are supervised algorithms that are used to build decision trees. The decision tree divides the dataset into smaller subsets according to information gain or the Gini index using decision nodes. To determine how effectively an attribute divides the training dataset according on its classification target, ID3 employs information gain. Depending on the values of the given attribute, information gain (IG) is the difference between the entropy of the dataset before and after splitting. Assume that $X = x_1, x_2, x_3, ...x_n$ is the set of instances, A is the attribute, and X_v is the subset of X where $A = v$. Information gain is then provided by

$$IG(X, A) = \text{Ent}(X) - \sum_{v \in V(A)} \frac{|X_v|}{|X|} \cdot \text{Ent}(X_v), \tag{5.13}$$

where $V(A)$ is the totality of all possible values for A, and ENT(X) is the entropy of X. Entropy of X is determined by:

$$\text{Ent}(X) = \sum_{i=1}^{c} -p_i \log_2 p_i, \tag{5.14}$$

where the probability of the current state X is indicated by p_i.

***K*-nearest neighbor:** The supervised machine learning method K-nearest neighbor (KNN) is non-parametric, which means it makes no assumptions about the underlying data. The KNN approach incorporates the undiscovered data in the class with the highest similarities to the existing data, ensuring that the undiscovered data and existing dataset are comparable. KNN acts by basically storing the data during training. When it encounters fresh data during testing, it determines the K-nearest neighbors to the most recent data using a distance metric, or Euclidean distance, and classifies it based on similarity [43]. Following are the steps of the KNN algorithm:

1. First, choose the nearest data points, or K value. The value of K can be any integer.
2. For each data point in the test data collection, perform the following tasks: (i) Determine the distance using the Manhattan, Euclidean, or Hamming distances between the data point and each sample in the training dataset. In this study, the Euclidean distance measure is used to calculate distance; (ii) samples are sorted in ascending order based on distance value; (iii) the top K samples are chosen as the closest neighbors to the test data point; and (iv) the test data point is then given a class based on the most prevalent class of these K samples.

Support vector machine: Support vector machine (SVM) is a method of data classification using supervised machine learning. SVM is a well-known method applied to numerous computational biology and bioinformatics issues. To describe the non-linear transition from the primary sequence to the protein structure area, SVM requires fewer model parameters. SVM will continuously create the hyperplane to reduce error. Fast training is a strength of SVM, which is important for high-throughput database testing [44]. Let (X_1, y_1), (X_2, y_2), (X_3, y_3),..., (X_n, y_n) be used to represent the dataset. SVM attempts to resolve the following equation:

$$\min_{w,b} ||w||^2 \text{such that } \forall i, y_i(\langle w, X_i \rangle + b) \geq 1. \tag{5.15}$$

where the weight and bias of the hyperplane equation $w \cdot X + b = 0$ are denoted by w and b.

5.3.3 Evaluation metrics

Since we have classified essential genes from five different species -AT, DOM, DR, HS, and MM–in this study, the essential genes classification problem is a multi-class classification problem. The evaluation matrices (accuracy, precision, recall, and F1-score) were computed for each class in the target. The final value of the assessment metrics was then determined using the weighted averaging technique.

$$\text{Accuracy} = \frac{\sum_{i=1}^{C} n_i \times \frac{TP_i+TN_i}{TP_i+TN_i+FP_i+FN_i}}{\sum_{i=1}^{C} n_i} \tag{5.16}$$

$$\text{Precision} = \frac{\sum_{i=1}^{C} n_i \times \frac{TP_i}{TP_i+FP_i}}{\sum_{i=1}^{C} n_i} \tag{5.17}$$

$$\text{Recall} = \frac{\sum_{i=1}^{C} n_i \times \frac{TP_i}{TP_i+FN_i}}{\sum_{i=1}^{C} n_i} \tag{5.18}$$

$$\text{F1} - \text{score} = \frac{\sum_{i=1}^{C} n_i \times \frac{2\times Precision_i \times Recall_i}{Precision_i+Recall_i}}{\sum_{i=1}^{C} n_i} \tag{5.19}$$

where

$$\text{Precision}_i = \frac{TP_i}{TP_i + FP_i} \tag{5.20}$$

and

$$\text{Recall}_i = \frac{TP_i}{TP_i + FN_i} \tag{5.21}$$

where the counts of true positives, true negatives, false positives, and false negatives for the ith class are TP_i, TN_i, FP_i, and FN_i. The number of classes in the problem is represented by C, and the number of samples in the ith class is represented by n_i.

5.3.4 Model framework

Based on co-occurrence matrices, the proposed approach categorized essential genes in five species. The proposed framework uses DNA sequences to extract eight separate co-occurrence matrices. Five features–energy, entropy, homogeneity, contrast, and dissimilarity–were taken from each co-occurrence matrix. The proposed features for the classification of essential genes were merged with the already-calculated features, such as Hurst exponent, fractal dimension, Shannon entropy, and modified Shannon entropy. The model was assessed using the SVM method, a supervised machine learning technique. The framework for the proposed model is depicted in Figure 5.1.

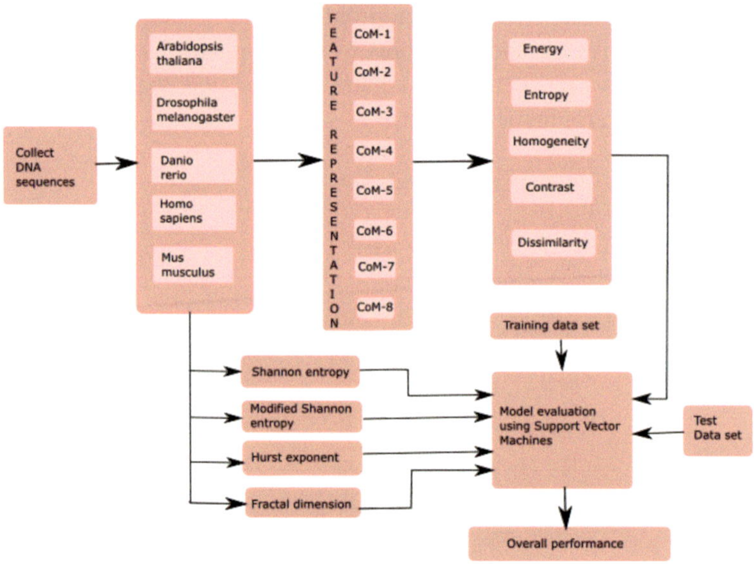

Figure 5.1 Framework of the proposed model for the classification of essential genes. Here, CoM indicates the co-occurrence matrices.

5.4 Result and Discussion

The proposed essential gene classification methodology can find novel essential genes with high recall and precision while only requiring a minimal number of previously recognized essential genes in some species. When examining essential genes in newly sequenced genomes of other species with few essential genes known, such a strategy could be incredibly helpful. The proposed study has been implemented in the "Python" environment, and data classification tasks have been carried out using the "Python" library of machine learning algorithms. The best scripting and programming language is Python, which is also open-source and includes sophisticated object-oriented ways for dealing with mathematical and statistical operations. The proposed methodology is put into practize in the Kaggle repository, which examines research for data scientists and machine learning engineers as the top experts in these domains. For data analysis and prediction models, we have used Python libraries like NumPy, Pandas, Matplotlib, SklearnPreprocessing, SklearnClassifiers, SklearnMetrics, and a few others. The *k-Nearest Neighbour (KNN), Decision Tree (DT), and Support Vector Machine (SVM)* classifiers are applied to the feature vectors extracted from each DNA gene sequence $\mathcal{S}$. Here are the datasets from the species AT, DOM, DR, HS, and

MM (see Table 5.12). There are several subsections in the experimenting of the proposed methodology.

Table 5.12 Demonstration of actual files containing gene sequences corresponds to AT, DOM, DR, HS, and MM species.

	Actual files	Actual files containing DNA sequences
AT	356	356
DOM	339	339
DR	315	315
HS	2054	2051
MM	411	125

5.4.1 Experiment for the proposed features

Individual feature tests have been carried out in this section. In this case, for the DNA sequence $\mathcal{S}$, individual features from each $f_I, f_J, f_K, f_L, f_M, f_N, f_O, f_P$ were considered, followed by a subsequent classification. The distribution of the F1-score performance for each of the 40 features derived from the co-occurrence matrices of the DNA sequence $\mathcal{S}$ is shown in Figure 5.2 for the decision-tree (DT), KNN, and SVM classifiers. As seen in this figure, both the KNN and SVM classifiers predict the classification problem more accurately than the decision tree (DT) classifier for most of the features. Additionally, it has been noted that classifiers have produced results that are roughly comparable for the most of features, with the exception of the 19th, 26th, 27th, 30th, 32nd, and 35th features of the 40-dimensional feature vector f, which have produced better results. The performance of KNN, DT, and SVM classifiers has been provided in Table 5.13 for assessing the effects of certain parameters, such as entropy, homogeneity, energy, contrast, and dissimilarity, on the classification of essential genes. The same training–testing techniques are used in all trials, and for each DNA sequence $\mathcal{S}$, the corresponding features are extracted from all co-occurrence matrices. Therefore, features for entropy, homogeneity, energy, contrast, and dissimilarity are recovered from each 8-dimensional feature vector.

From Table 5.13, it can be seen that performance for each feature is essentially the same, although the KNN classifier performs better than DT and SVM. The employed species AT, DOM, DR, HS, and MM have issues with class imbalance; hence, the F1-score has been taken into consideration here as classification performance. Additionally, it has been thought about how the features derived from each co-occurrence matrix will affect the

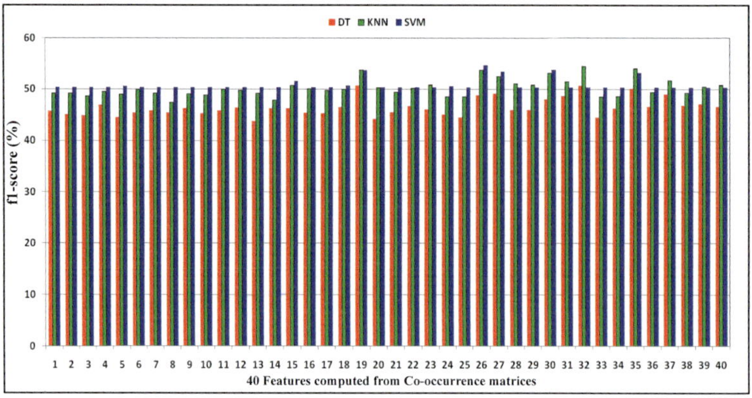

Figure 5.2 Demonstration of distribution of F1-score performance obtained by decision-tree, KNN, and SVM classifiers with respect to each of the 40 features computed from co-occurrence matrices of DNA gene sequence S.

Table 5.13 Impact of different co-occurrence features on the classification of essential gene sequences of AT, DOM, DR, HS, and MM species.

Classifier	Accuracy	Precision	Recall	F1-score
Effect of entropy features				
K-nearest-neighbors	63.56	56.68	63.56	59.39
Decision tree	52.95	53.56	52.95	53.25
Support vector machine	64.37	41.44	64.37	50.42
Effect of dissimilarity features				
K-nearest-neighbors	62.96	57.38	62.96	59.55
decision tree	52.70	53.84	52.70	53.25
Support vector machine	67.07	58.80	67.07	56.75
Effect of energy features				
K-nearest neighbors	59.48	52.71	59.48	55.46
Decision tree	48.65	49.82	48.65	49.22
Support vector machine	64.94	50.32	64.94	51.83
Effect of homogeneity features				
K-nearest-neighbors	63.06	57.59	63.06	59.99
Decision tree	53.61	54.81	53.61	54.19
Support vector machine	67.67	60.76	67.67	58.29
Effect of contrast features				
K-nearest-neighbors	64.25	58.92	64.25	61.02
Decision tree	54.80	56.27	54.80	55.51
Support vector machine	68.36	59.82	68.36	58.85

succeeding experiments. Each co-occurrence matrix is now used to extract the five-dimensional feature vector. Under the identical training–testing strategy, the performance related to these feature vectors is reported in Table 5.14. Table 5.14 shows that the co-occurrence matrix features have an effect on the essential gene classification that is more or less similar. The features generated from the co-occurrence matrices are therefore beneficial and efficient. KNN classifier performs better in this instance as well.

Table 5.14 Impact of features extracted from different co-occurrence matrices for the classification of essential gene sequences of AT, DOM, DR, HS, and MM species.

Classifier	Accuracy	Precision	Recall	F1-score
Effect of first matrix				
K-nearest-neighbors	63.37	56.39	63.37	59.20
Decision tree	53.70	54.02	53.70	53.85
Support vector machine	64.38	41.44	64.38	50.42
Effect of second matrix				
K-nearest neighbors	62.05	54.43	62.05	57.54
Decision tree	53.20	53.88	53.20	53.53
Support vector machine	64.38	41.44	64.38	50.42
Effect of third matrix				
K-nearest neighbors	60.58	52.69	60.58	55.66
Decision tree	49.72	51.01	49.72	50.34
Support vector machine	64.38	41.44	64.38	50.42
Effect of fourth matrix				
K-nearest-neighbors	62.96	58.32	62.96	59.41
Decision tree	54.33	55.14	54.33	54.72
Support vector machine	64.38	41.44	64.38	50.42
Effect of fifth matrix				
K-nearest neighbors	57.91	49.72	57.91	53.02
Decision-Tree	47.24	48.14	47.24	47.69
Support vector machine	64.38	41.44	64.38	50.42
Effect of sixth matrix				
K-nearest neighbors	61.49	54.13	61.49	57.14
Decision tree	52.69	54.34	52.69	53.49
Support vector machine	65.35	47.61	65.35	53.36
Effect of seventh matrix				
K-Nearest neighbors	58.82	52.94	58.82	55.37
Decision tree	50.44	51.56	50.44	50.99
Support vector machine	64.81	46.81	64.81	53.45
Effect of eighth matrix				
K-nearest neighbors	56.12	50.86	56.12	52.78
Decision tree	49.28	49.86	49.28	49.56
Support vector machine	64.38	41.44	64.38	50.42

5.4.2 Experiment for the existing features

In a subsequent experiment, a performance comparison was done using various relevant state-of-the-art feature extraction approaches, where these features are extracted appropriately. These techniques include Hurst exponent, modified Shannon entropy, Shannon entropy, and fractal dimension. With regard to KNN, DT, and SVM classifiers, the performance is attained. Table 5.15 reports the performance resulting from these features. According to this table, the performance of the Hurst exponent, modified Shannon entropy, fractal dimension characteristics, and Shannon entropy are all quite comparable. SVM has, nevertheless, performed better than the other classifiers. The proposed technique has better categorized the essential genes of AT, DOM, DR, HS, and MM species under the identical training–testing protocol, as evidenced by the comparison of these performances with the proposed system in Figure 5.3. The 40-dimensional feature vector is taken into account in the proposed system, whereas each of the currently used

Table 5.15 Impact of existing and proposed features on the classification of essential genes for the AT, DOM, DR, HS, and MM species.

Classifier	Accuracy	Precision	Recall	F1–score
Effect of Shannon entropy features				
K-nearest neighbors	53.10	46.24	53.10	49.14
Decision tree	48.28	46.96	48.28	47.53
Support vector machine	64.33	41.38	64.33	50.36
Effect of Hurst exponent features				
K-nearest neighbors	53.98	45.63	53.98	49.14
Decision tree	43.57	45.41	43.57	44.45
Support vector machine	64.33	41.38	64.33	50.36
Effect of modified Shannon entropy features				
K-nearest neighbors	54.67	46.20	54.67	49.71
Decision tree	41.76	43.98	41.76	42.80
Support vector machine	64.26	45.64	64.26	50.66
Effect of fractal dimension features				
K-nearest neighbors	58.11	52.19	58.11	52.15
Decision tree	68.35	46.72	68.35	55.51
Support vector machine	68.35	46.72	68.35	55.51
Effect of proposed features				
K-nearest neighbors	64.95	59.49	64.95	61.50
Decision tree	58.31	59.24	58.31	58.70
Support vector machine	66.14	56.57	66.14	54.35

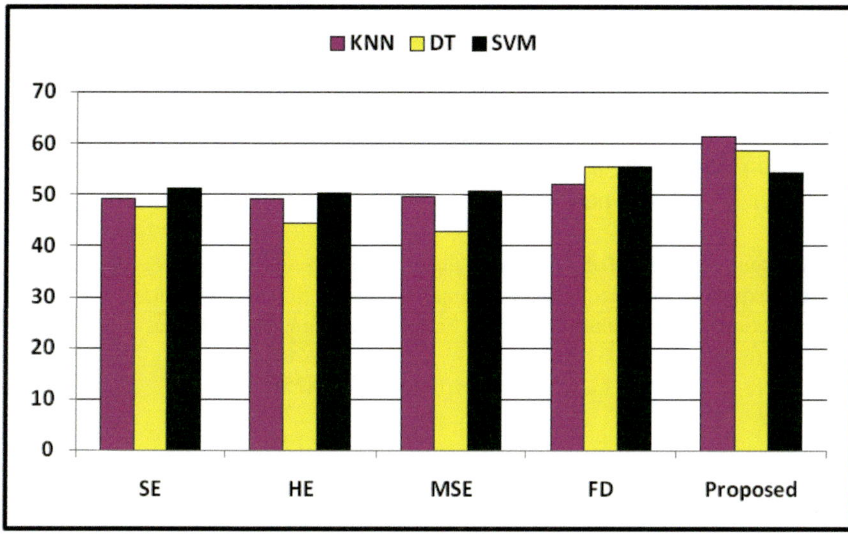

Figure 5.3 Performance (F1-score) comparison of existing features and the proposed features for the classification of essential genes of AT, DOM, DR, HS, and MM species.

feature extraction techniques only extracts one-dimensional feature vectors. With superior performance than the current state-of-the-art features, this work investigates the discriminatory potential of co-occurrence matrix features.

5.4.3 Experiment for the combined features

The distribution of these nucleotides and information on their relative positions within the gene S are derived from the co-occurrence of the nucleotides A, C, T, G in the essential gene. Modern feature extraction methods, which are presented in this article, are key concepts in information theory. For instance, the nucleotide uncertainty and randomness in the gene S are calculated using Shannon entropy and its modified method. For the purpose of analyzing its distribution in the essential gene, the Hurst exponent evaluates the relative tendency and distinctive parameters. From the indicator matrix derived from the essential gene S, the fractal dimension estimates the distribution of nucleotides that matches a fractal. In order to obtain more discriminant features for the classification of essential genes of AT, DOM, DR, HS, and MM species, randomness, relative tendency, and fractal-like distribution information in S are combined here based on the similarity of patterns of nucleotides computed by the co-occurrence matrices and the information of uncertainty. To determine the most efficient fusion of these features, the principal component analysis of dimensionality reduction with variation ratio has been used. Table 5.16 shows how well these features work together in terms of performance. In relation to different dimensionally

Table 5.16 Demonstration of discriminant features among proposed features, Shannon-entropy, Hurst exponent, modified Shannon entropy, and fractal dimension features.

Feature	Eigen values	Rank	Feature	Eigen values	Rank
f_1	13.908	1	f_{23}	0.283	23
f_2	4.434	2	f_{24}	0.257	24
f_3	3.628	3	f_{25}	0.224	25
f_4	2.895	4	f_{26}	0.192	26
f_5	2.505	5	f_{27}	0.152	27
f_6	2.233	6	f_{28}	0.109	28
f_7	1.904	7	f_{29}	0.041	29
f_8	1.602	8	f_{30}	0.032	30
f_9	1.388	9	f_{31}	0.027	32
f_{10}	1.133	10	f_{32}	0.027	31
f_{11}	0.986	11	f_{33}	0.023	33
f_{12}	0.855	12	f_{34}	0.019	34
f_{13}	0.820	13	f_{35}	0.015	35
f_{14}	0.750	14	f_{36}	0.008	36
f_{15}	0.714	15	f_{37}	0.006	37
f_{16}	0.525	16	f_{38}	0.001	43
f_{17}	0.471	17	f_{39}	0.001	44
f_{18}	0.440	18	f_{40}	0.002	42
f_{19}	0.432	19	f_{41}	0.003	41
f_{20}	0.333	20	f_{42}	0.003	40
f_{21}	0.329	21	f_{43}	0.004	39
f_{22}	0.299	22	f_{44}	0.004	38

reduced features, Table 5.17 shows the discriminatory strength of combined features for KNN, DT, and SVM classifiers. This table shows that the SVM classifier is responsible for the highest F1-score, which is 71.42. F1-score performance has been reported because of the class imbalance issue.

A final performance for the combination of features for classification of essential genes of AT, DOM, DR, HS, and MM species has been displayed in Figure 5.4 for simpler interpretation and visibility.

Table 5.17 Demonstration of performance due to the combination of features for the classification of essential genes of AT, DOM, DR, HS, and MM species.

Variation	Classifier	Accuracy	Precision	Recall	F1-score	Feature dimension
0.85	*K*-nearest neighbors	72.01	66.37	72.01	68.67	4
	Decision tree	63.09	63.63	63.09	63.34	
	Support vector machine	74.30	68.77	74.30	67.69	
0.9	*K*-nearest neighbors	71.52	66.77	71.52	68.94	5
	Decision tree	62.67	63.81	62.67	63.18	
	Support vector machine	75.91	69.57	75.91	70.31	
0.95	*K*-nearest neighbors	73.82	68.83	73.82	70.80	7
	Decision tree	63.93	64.67	63.93	64.29	
	Support vector machine	76.46	72.63	76.46	71.06	
0.99	*K*-nearest neighbors	73.96	68.29	73.96	70.66	9
	Decision tree	64.48	65.35	64.48	64.88	
	Support vector machine	76.32	70.56	76.32	**71.42**	

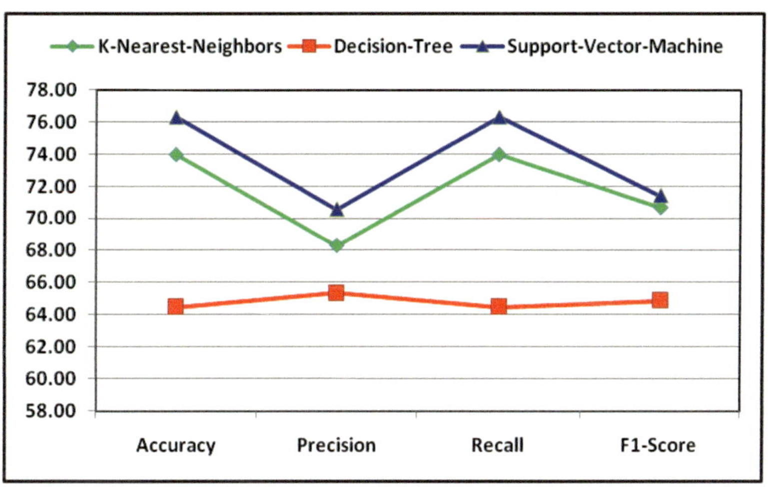

Figure 5.4 Demonstration of final performance for the combination of features for the classification of essential genes of AT, DOM, DR, HS, and MM species.

5.5 Conclusion

In this work, a novel feature extraction and analysis method has been taken into consideration for the classification of essential genes in the species *Arabidopsis thaliana* (AT), *Drosophila melanogaster* (DOM), *Danio rerio* (DR), *Homo sapiens* (HS), and *Mus musculus* (MM). The planned scheme is to be implemented in three stages. The distribution of nucleotides and their relative positions are derived from each gene, in the first segment, using unique co-occurrence matrix-based features. The statistical analysis of the distribution of stationary patterns of nucleotides in essential genes uses the features from these metrics. In the second section, information theory measures that compute uncertainty, unpredictability, relative tendency, and fractal-like patterns in the gene are applied to certain current state-of-the-art feature computation approaches, such as Shannon entropy, Hurst exponent, and fractal dimension. The features from the proposed methodology and the existing methodologies are subjected to individual classification tasks in the third section of this study, where their F1-score performance has been taken into consideration for comparison. These comparisons demonstrate the reliability and potency of the proposed methodology. Finally, more discriminatory features are computed for classifying essential genes of AT, DOM, DR, HS, and MM species using the features from the proposed approach and the currently used methodologies.

References

[1] Juhas, Mario and Eberl, Leo and Glass, John I, Essence of life: essential genes of minimal genomes, Trends in cell biology, volume 21, number 10, pages 562–568, 2011, Elsevier.

[2] Koonin, Eugene V, How many genes can make a cell: the minimal-gene-set concept, Annual review of genomics and human genetics, volume 1, number 1, pages 99-116, 2000, Annual Reviews 4139 El Camino Way, PO Box 10139, Palo Alto, CA 94303-0139, USA.

[3] Umer, Saiyed and Dhara, Bibhas Chandra and Chanda, Bhabatosh, Texture code matrix-based multi-instance iris recognition, Journal of Pattern Analysis and Applications, volume 19, number 1, pages 283-295, 2016, Springer.

[4] Le, Nguyen Quoc Khanh and Do, Duyen Thi and Hung, Truong Nguyen Khanh and Lam, Luu Ho Thanh and Huynh, Tuan-Tu and Nguyen, Ngan Thi Kim, A computational framework based on ensemble deep

neural networks for essential genes identification, International journal of molecular sciences, volume 21, number 23, pages 9070, 2020, MDPI.

[5] Juhas, Mario and Reuß, Daniel R and Zhu, Bingyao and Commichau, Fabian M, Bacillus subtilis and Escherichia coli essential genes and minimal cell factories after one decade of genome engineering, Microbiology, volume 160, number 11, pages 2341–2351, 2014, Microbiology Society.

[6] Itaya, Mitsuhiro, An estimation of minimal genome size required for life, FEBS letters, volume 362, number 3, pages 257-260, 1995, Elsevier.

[7] Dickerson, Jonathan E and Zhu, Ana and Robertson, David L and Hentges, Kathryn E, Defining the role of essential genes in human disease, PloS one, volume 6, number 11, pages e27368, 2011, Public Library of Science San Francisco, USA.

[8] Qin, Hong, Estimating network changes from lifespan measurements using a parsimonious gene network model of cellular aging, Bmc Bioinformatics, volume 20, number 1, pages 1-10, 2019, BioMed Central.

[9] Guo, Hao-Bo and Ghafari, Mehran and Dang, Weiwei and Qin, Hong, Protein interaction potential landscapes for yeast replicative aging, Scientific reports, volume 11, number 1, pages 1-12, 2021, Nature Publishing Group.

[10] Roemer, Terry and Jiang, Bo and Davison, John and Ketela, Troy and Veillette, Karynn and Breton, Anouk and Tandia, Fatou and Linteau, Annie and Sillaots, Susan and Marta, Catarina and others, Large-scale essential gene identification in Candida albicans and applications to antifungal drug discovery, Molecular microbiology, volume 50, number 1, pages 167-181, 2003, Wiley Online Library.

[11] Cullen, Lara M and Arndt, Greg M, Genome-wide screening for gene function using RNAi in mammalian cells, Immunology and cell biology, volume 83, number 3, pages 217-223, 2005, Wiley Online Library.

[12] Veeranagouda, Yaligara and Husain, Fasahath and Tenorio, Elizabeth L and Wexler, Hannah M, Identification of genes required for the survival of B. fragilis using massive parallel sequencing of a saturated transposon mutant library, BMC genomics, volume 15, number 1, pages 1-11, 2014, Springer.

[13] Giaever, Guri and Chu, Angela M and Ni, Li and Connelly, Carla and Riles, Linda and Véronneau, Steeve and Dow, Sally and Lucau-Danila, Ankuta and Anderson, Keith and André, Bruno and others, Functional

profiling of the Saccharomyces cerevisiae genome, nature, volume 418, number 6896, pages 387-391, 2002, Nature Publishing Group.

[14] Khandelwal, Monika and Rout, Ranjeet Kumar and Umer, Saiyed, Protein-protein interaction prediction from primary sequences using supervised machine learning algorithm, IEEE 12th International Conference on Cloud Computing, Data Science & Engineering (Confluence), pages 268-272, 2022, IEEE.

[15] Khandelwal, Monika and Sheikh, Sabha and Rout, Ranjeet Kumar and Umer, Saiyed and Mallik, Saurav and Zhao, Zhongming, Unsupervised Learning for Feature Representation Using Spatial Distribution of Amino Acids in Aldehyde Dehydrogenase (ALDH2) Protein Sequences, Mathematics, volume 10, number 13, pages 2228, 2022, Multidisciplinary Digital Publishing Institute (MDPI).

[16] Khandelwal, Monika and Kumar Rout, Ranjeet and Umer, Saiyed and Mallik, Saurav and Li, Aimin, Multifactorial feature extraction and site prognosis model for protein methylation data, Briefings in Functional Genomics, 2022.

[17] Mobegi, Fredrick M and Zomer, Aldert and De Jonge, Marien I and Van Hijum, Sacha AFT, Advances and perspectives in computational prediction of microbial gene essentiality, Briefings in functional genomics, volume 16, number 2, pages 70-79, 2017, Oxford University Press.

[18] Juhas, Mario and Stark, Manuel and von Mering, Christian and Lumjiaktase, Puthapoom and Crook, Derrick W and Valvano, Miguel A and Eberl, Leo, High confidence prediction of essential genes in Burkholderia cenocepacia, PloS one, volume 7, number 6, pages e40064, 2012, Public Library of Science San Francisco, USA.

[19] McCutcheon, John P and Moran, Nancy A, Functional convergence in reduced genomes of bacterial symbionts spanning 200 My of evolution, Genome biology and evolution, volume 2, pages 708-718, 2010, Oxford Academic.

[20] Gil, Rosario and Silva, Francisco J and Peretó, Juli and Moya, Andrés, Determination of the core of a minimal bacterial gene set, Microbiology and Molecular Biology Reviews, volume 68, number 3, pages 518-537, 2004, Am Soc Microbiol.

[21] Chen, Yu and Xu, Dong, Understanding protein dispensability through machine-learning analysis of high-throughput data, Bioinformatics, volume 21, number 5, pages 575-581, 2005, Oxford University Press.

[22] Seringhaus, Michael and Paccanaro, Alberto and Borneman, Anthony and Snyder, Michael and Gerstein, Mark, Predicting essential genes in

fungal genomes, Genome research, volume 16, number 9, pages 1126-1135, 2006, Cold Spring Harbor Lab.

[23] Yuan, Yuan and Xu, Yanxun and Xu, Jianfeng and Ball, Robyn L and Liang, Han, Predicting the lethal phenotype of the knockout mouse by integrating comprehensive genomic data, Bioinformatics, volume 28, number 9, pages 1246-1252, 2012, Oxford University Press.

[24] Deng, Jingyuan, An integrated machine-learning model to predict prokaryotic essential genes, Gene Essentiality, pages 137-151, 2015, Springer.

[25] Zhang, Xue and Xiao, Wangxin and Xiao, Weijia, DeepHE: Accurately predicting human essential genes based on deep learning, PLoS Computational Biology, volume 16, number 9, pages e1008229, 2020, Public Library of Science San Francisco, CA USA.

[26] Liu, Xiao and Wang, Bao-Jin and Xu, Luo and Tang, Hong-Ling and Xu, Guo-Qing, Selection of key sequence-based features for prediction of essential genes in 31 diverse bacterial species, PLoS One, volume 12, number 3, pages e0174638, 2017, Public Library of Science San Francisco, CA USA.

[27] Rout, Ranjeet Kumar and Hassan, SK Sarif and Sindhwani, Sanchit and Pandey, Hari Mohan and Umer, Saiyed, Intelligent classification and analysis of essential genes using quantitative methods, ACM Transactions on Multimedia Computing, Communications, and Applications (TOMM), volume 16, number 1s, pages 1-21, 2020, ACM New York, NY, USA.

[28] Xu, Luo and Guo, Zhirui and Liu, Xiao, Prediction of essential genes in prokaryote based on artificial neural network, Genes & genomics, volume 42, pages 97-106, 2020, Springer.

[29] Chen, Hebing and Zhang, Zhuo and Jiang, Shuai and Li, Ruijiang and Li, Wanying and Zhao, Chenghui and Hong, Hao and Huang, Xin and Li, Hao and Bo, Xiaochen, New insights on human essential genes based on integrated analysis and the construction of the HEGIAP web-based platform, Briefings in Bioinformatics, volume 21, number 4, pages 1397-1410, 2020, Oxford University Press.

[30] Kuang, Shuzhen and Wei, Yanzhang and Wang, Liangjiang, Expression-based prediction of human essential genes and candidate lncRNAs in cancer cells, Bioinformatics, volume 37, number 3, pages 396-403, 2021, Oxford University Press.

[31] Senthamizhan, Vimaladhasan and Ravindran, Balaraman and Raman, Karthik, NetGenes: A database of essential genes predicted using

features from interaction networks, Frontiers in Genetics, volume 12, pages 722198, 2021, Frontiers Media SA.

[32] Marques de Castro, Giovanni and Hastenreiter, Zandora and Silva Monteiro, Thiago Augusto and Martins da Silva, Thieres Tayroni and Pereira Lobo, Francisco, Cross-species prediction of essential genes in insects, Bioinformatics, volume 38, number 6, pages 1504-1513, 2022, Oxford University Press.

[33] Zurek, Wojciech H, Algorithmic randomness and physical entropy, Physical Review A, volume 40, number 8, pages 4731, 1989, APS.

[34] Hurst, Harold Edwin, Long-term storage capacity of reservoirs, Transactions of the American society of civil engineers, volume 116, number 1, pages 770-799, 1951, American Society of Civil Engineers.

[35] Hassan, Sk Sarif and Rout, Ranjeet Kumar and Sahoo, Kshira Sagar and Jhanjhi, Nz and Umer, Saiyed and Tabbakh, Thamer A and Almusaylim, Zahrah A, A Vicenary Analysis of SARS-CoV-2 Genomes, Cmc-Computers Materials & Continua, pages 3477-3493, 2021.

[36] Rout, Ranjeet Kumar and Hassan, Sk Sarif and Sheikh, Sabha and Umer, Saiyed and Sahoo, Kshira Sagar and Gandomi, Amir H, Feature-extraction and analysis based on spatial distribution of amino acids for SARS-CoV-2 Protein sequences, Computers in Biology and Medicine, volume 141, pages 105024, 2022, Elsevier.

[37] Rout, Ranjeet Kumar and Pal Choudhury, Pabitra , and Maity, Santi Prasad and Daya Sagar, BS and Hassan, Sk Sarif, Fractal and mathematical morphology in intricate comparison between tertiary protein structures, Computer Methods in Biomechanics and Biomedical Engineering: Imaging & Visualization, volume 6, number 2, pages 192-203, 2018, Taylor & Francis.

[38] Umer, Saiyed and Mohanta, Partha Pratim and Rout, Ranjeet Kumar and Pandey, Hari Mohan, Machine learning method for cosmetic product recognition: a visual searching approach, Multimedia Tools and Applications, volume 80, number 28, pages 34997-35023, 2021, Springer.

[39] Cattani, Carlo, Fractals and hidden symmetries in DNA, Mathematical problems in engineering, volume 2010, 2010, Hindawi.

[40] Upadhayay, Pranav Dutt and Agarwal, Rupaansh Chandra and Rout, Ranjeet Kumar and Agrawal, Arun Prakash, Mathematical Characterization of Membrane Protein Sequences of Homo-Sapiens, 2019 9th International Conference on Cloud Computing, Data Science & Engineering (Confluence), pages 382-386, 2019, IEEE.

[41] Rout, Ranjeet Kumar and Ghosh, Suvankar and Choudhury, Pabitra Pal, Classification of Mer Proteins in a Quantitative Manner, InternationalJournal of Computer Applications in Engineering Science, volume 4, number 2, pages 31-34, 2014.

[42] Quinlan, J. Ross, Induction of decision trees, Machine learning, volume 1, number 1, pages 81-106, 1986, Springer.

[43] Peterson, L. E. (2009). K-nearest neighbor. Scholarpedia 4, 1883

[44] Suthaharan, Shan, Support vector machine, book: Machine learning models and algorithms for big data classification, pages 207-235, 2016, Springer.

[45] Rout RK, Umer S, Khandelwal M, Pati S, Mallik S, Balabantaray BK, Qin H. Identification of discriminant features from stationary pattern of nucleotide bases and their application to essential gene classification. Front Genet. 2023 Apr 20;14:1154120. doi: 10.3389/fgene.2023.1154120. PMID: 37152988; PMCID: PMC10156977.

6

Integrating Deep Learning and Next-Generation Medical Image Data for Rare Disease Stage Detection

6.1 Introduction

The second-leading cause of death in recent years, behind cardiovascular disease, is cancer, which is accounting for approximately 10 million mortality, according to the World Health organization (WHO) World Cancer Report (2020) [34]. Cancer screening requires different and more complex public health strategies than other diagnoses, as well as more infrastructure, coordination, and money. When the following criteria are met, WHO advises the implementation of routine screening [34]:

1. It has been shown that the tool, model, and software are effective.
2. There are enough facilities and resources to verify diagnosis and therapies.
3. The severe illness prevalence justifies the screening.

Globally, there are 3.9 central nervous system tumors for every 100,000 people; the incidence varies by age, gender, race, and area, and is highest in Northern Europe, then Australia, the USA, and Canada. Meningioma is the most prevalent malignant tumor, accounting for 75% of central nervous system malignant tumors, and accounts for 36.8% of all tumors. It has an annual incidence of six cases per 100,000 persons. Due to its non-invasive nature, MRI is currently the best approach for the early detection of human brain tumors [33]. Nonetheless, the opinions of radiologists have a major role in how an MRI is interpreted.

Imaging-based artificial intelligence (AI) solutions have their roots in the development of convolution neural network (CNN) based deep learning (DL). In order to support clinical decision-making, DL-guided solutions are designed. The CNN-based DL architecture that is being proposed does so for a variety of causes. First, it is noted that CNN-based DL succeeds at reducing the parameter threshold while preserving model quality. Second, since it can automatically extract features from an image, it does not need human feature engineering. Third, the CNN-based DL model, which was developed by a number of researchers and obtained good image classification and recognition accuracy, is supported by the literature. It is important to note that relatively few studies have integrated CNN with Shapley additive explanation (SHAP) and local interpretable model-agnostic explanation (LIME). Researchers showed the enormous potential of imaging techniques to lessen the onerous workload on medical professionals [37]. Also, it permits allocating more assistance to patient care, lowering burnout, and decreasing the overall medical costs of patients [9]. In order to accomplish a fully autonomous and quick diagnosis, authors in [14] employed DL algorithms, ResNet50, to identify COVID-19 from X-rays. The authors intended to reduce the run time to about 2.5 seconds, with an average COVID-19 detection time of roughly 2.5 seconds and an average accuracy of 0.97. The authors in [21] introduced several performance metrics, including DL, F1 value, precision, responsiveness, and specificity. The findings demonstrated typical sensitivity of 94.79% and accuracy of 92.93% for robust identification and detection of COVID-19 in the chest X-ray dataset. In one study [18], CNN architecture palettes were implemented to modify the deep neural network correlation learning mechanism for CT brain tumor detection to achieve the best feasible ANN detection result.

The AISA framework for MRI data processing showed how it might be used to brain scan data by generating independent subspaces and reducing textural information. Then, dimensionality is diminished for discriminative classification using t-SNE embedding. The KNN classifier is then used. The largest problem, particularly in medical settings, is that algorithm-driven conclusions are not transparent or interpretable, despite the enormous use of DL models in clinical decision-making. However, a number of researchers [28, 39] found that the lack of a standard view of explainability created a number of barriers to the development of XAI-based clinical decision support systems (CDSS). Using a dataset of MRI images, our study suggests an explanation-driven DL-based model to identify several subtypes of brain tumors (meningioma, glioma, and pituitary). To increase model transparency while maintaining a high performance rate, we additionally used LIME and Shapley additive explanations. Medical experts, physicians, and other users will be benefited by the assistance of our study in understanding and effectively managing the increasing number of trustworthy and dependable AI partners [31].

Our model, as compared to earlier models, used a dual-input CNN strategy to overcome the classification difficulty with lower-quality images and an accuracy of 94.64% when compared to other state-of-the-art models. Since previous research have been difficult to understand, we used explainable AI (XAI) algorithms like LIME and SHAP in this study, which sets it apart from others. As Shapley values check all potential combinations of inputs, we employed SHAP to guarantee consistency and local accuracy for interpretation. On the other hand, LIME builds sparse linear models around each prediction to explain how the model operates effectively.

In order to optimize the CNN design for the best potential detection result of DL, the computed tomography (CT) brain tumor detection method applied deep neural network correlation learning. Although the earlier models were more accurate, they lacked transparency, explainability, and interpretability [1, 19]. The suggested model applied XAI techniques like LIME [35] and SHAP, which were described in an algorithm. Deep neural network correlation learning mechanism for computed tomography (CT) brain tumor detection used palettes of CNN architecture to adjust them to the best possible detection result of DL. Though the previously suggested models have higher accuracy, they lack explainability, interpretability, and transparency [1, 19]. The proposed model used XAI algorithms like LIME [35] and SHAP as detailed on an algorithm 3.

The following is a summary of the contributions in this study [13]:

1. In order to provide a thorough explanation of the results, SHAP and LIME were employed to build a multi-input DL model that was explanation-driven. The convolution layer receives one set of the two input datasets, and the fully connected layer receives the other set.
2. Contrasting our brain MRI accuracy to other state-of-the-art methods, we have a high accuracy score of 94.64%.

6.2 Materials and Methods

6.2.1 Datasets

We used the freely downloadable MRI images for this study [7]. Along with the normal image, the datasets are annotated with three different types of tumors: glioma, meningioma, and pituitary. Out of a total of 2870 images,

2296 images of various categories are utilized as training sets, while the remaining 574 images are used as test sets.

6.2.1.1 Data Preprocessing

The size of each 512 × 512 × 3 image is reduced to 150 × 150 × 3. In order to achieve faster convergence, the images are shuffled, which stops the CNN model from learning the training order. We added Gaussian noise to the classification process because it enhances the learning for DL [25] with mean = 0, standard deviation as $10^{0.5}$. In the classification of tumors from the dataset, one case is shown in Figure 6.1.

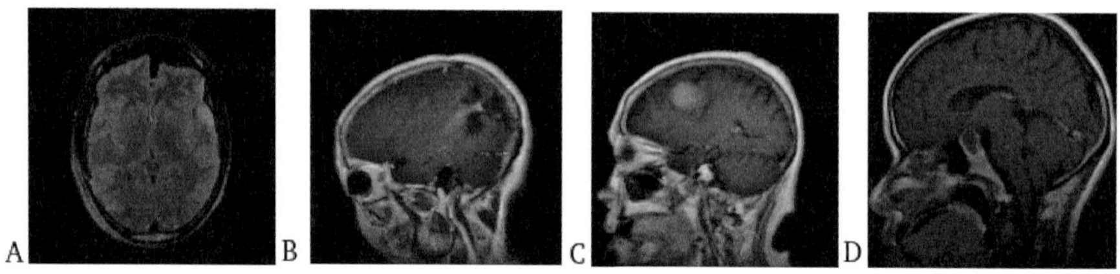

Figure 6.1 Sample image data of different types of tumors. (A) Normal: The intensity of the parenchyma in the brain without any tumor is normal. The ventricular system and cisternal spaces are supposed to be in good working order. There is always no evidence of an intracranial space-occupying lesion [11]. (B) Glioma tumor: Gliomas have thick, irregularly enhancing borders a focal necrotic core with a hemorrhagic component. They are surrounded by vasogenic-type edema, containing malignant cell infiltration. Rarely (less than 2%) intra-tumoral hemorrhage happens [10]. (C) Meningioma tumor: Meningiomas are extra-axial tumors arising from meningocytes or arachnoid cap cells of meninges and can be found where meninges exist, as well as in some sites where only rest cells are thought to exist [12]. (D) Pituitary tumor: For pituitary adenomas, minor intra-pituitary lesions appear differently than larger lesions that spread into the suprasellar region and pose various surgical and diagnostic issues. Based on tumor aspects, overall signal qualities can vary [36].

6.2.2 Proposed Framework

Figure 6.2 depicts the general architecture of the model, which consists of frameworks for feature extraction, the CNN model, statistical performance metrics, and explanation extraction. Two copies of the dataset are given to the CNN model with an output layer that is 1 × 4 in size and six hidden layers in order to increase accuracy [38]. The rectified linear unit (ReLU) and softmax are used as the activation function in the Adam optimizer while its default parameters are used. LIME and SHAP are used to test the statistical

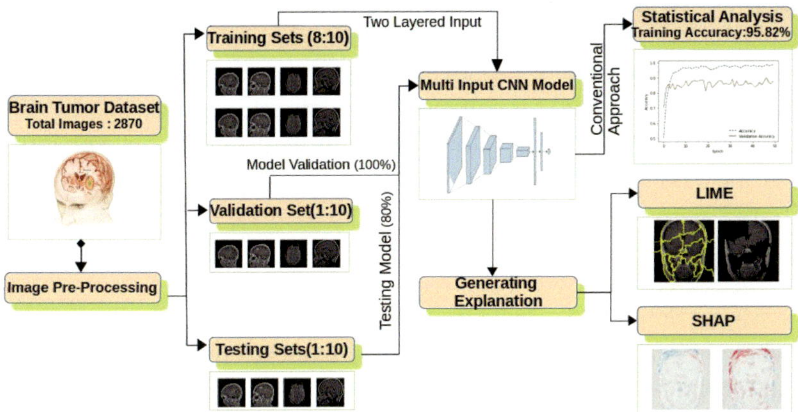

Figure 6.2 The proposed explanation-driven DL model for prediction of brain tumor status using MRI image data: 2870 MRI images are preprocessed and divided into training, validation, and test sets. Two copies of datasets are fed into a multi-input CNN model to find the training, validation, and test accuracy. The same CNN model was further imposed on LIME and SHAP.

accuracy of the final CNN model. While a gradient explainer is used for SHAP explanations, perturbation is calculated for LIME explanations. The algorithm 2 formalizes the entire procedure.

Algorithm 2 Explanation-driven multi-input DL model for prediction of brain tumor

Input: MRI Dataset (Breakdown in ratio 8:1:1) with size $150 \times 150 \times 3$
1: **Epoch:** 20
2: **Optimizer:** Adam
3: **Kernel Size:** 3×3
4: **Dropout:** 0.2
5: **Filter :** Conv2D
6: **for** *For every iteration in dual input CNN Model* **do**
7: Input one set of dataset to convolution layer
8: Input one set of dataset to fully connected layer
9: Calculate loss, accuracy, validation loss, validation accuracy
10: **end for**
11: **Implement XAI:** Implement LIME and SHAP for the model

The proposed explainable model employs CNN with dual input architecture for the classification work. In all hidden layers, CNN is enforced using ReLU as activation. The calculation of ReLU is straightforward when compared to the input value and zero value. In addition, ReLU has a derivative that depends on the positive or negative input and can be either 0 or 1. This ReLU functionality is crucial for contrasting explainable modules like LIME and SHAP. The Adam optimizer is applied with its default parameter [20] along with sparse categorical cross-entropy; the kernel size is set to 3×3.

6.3 Results

After classifying data, CNN model performance is assessed based on accuracy and the number of incorrect predictions. Figure 6.3 presents the curves for conventional outcomes of CNN.

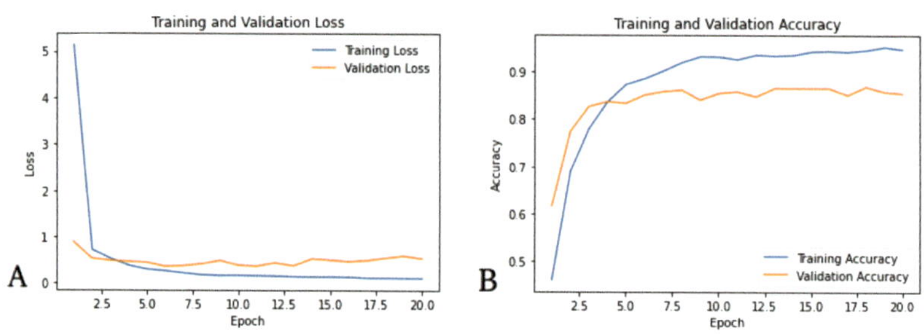

Figure 6.3 Training and validation results of CNN.

6.3.1 CNN

The model was iterated for 20 epochs, and we had observed the loss with min mode and patience level of 3 to cross the over-fitting during callback in CNN modules. The model contains 26 incorrect predictions with a training loss of 0.1149 and a validation loss of 0.53 while achieving an overall test accuracy of 85.37% and a training accuracy of 94.64%, respectively.

In order to further evaluate the performance of the CNN model on the specified dataset, K-fold cross-validation is carried out with $K = 10$ non-overlapping folds for 20 epochs with a batch size of 128. The test and training sets were divided into 1:4. Table 6.1 displays the final validation result of the cross-fold. Almost 100% training accuracy was obtained by the proposed model during cross-validation.

The confusion matrix for 287 test images is shown in Table 6.2. Wrong classifications were made for seven normal images out of 46, 14 glioma

Table 6.1 K-fold cross-validation results.

Fold	Final validation loss	Final validation accuracy (%)
1	0.01144	99.5
2	0.01706	98.47
3	0.02152	99.13
4	0.00988	99.34
5	0.00554	100
6	0.01298	99.13
7	0.00874	99.78
8	0.00533	99.78
9	0.01018	99.34
10	0.0088	99.56

images out of 84, 12 meningioma images out of 77, and 3 pituitary images out of 80.

Table 6.2 Confusion matrix for CNN.

		Actual value			
		Normal	Glioma	Meningioma	Pituitary
Predicted values	Normal	37	8	1	0
	Glioma	7	70	5	2
	Meningioma	0	12	65	0
	Pituitary	0	3	0	77

We used the McNemar test [32] to statistically validate our model. It provided a chi-squared value of 42.022 and a p-value of $9.02 \times e^{-11}$ for the labels of the test data and labels of the prediction model under the test data. Since the p-value is less than $\alpha = 0.005$, we can rule out the null hypothesis that both labels perform equally well on the test set.

6.3.2 SHAP

The scores indicate the contribution of each pixel on a projected image and can be used to explain tumor classification tasks. According to the algorithm 3, the Shapley values correspond to each feature for various categories of the tumor.

Direct interpretation of a CNN model with mathematical behavior is challenging. Therefore, using SHAP, it is easy to understand how each input feature affects the output of the model, which is displayed in Figures 6.4 and 6.5. Red pixels show positive SHAP values that boost the probability of the class. Blue pixels are used to signify negative SHAP values that reduce

Algorithm 3 Algorithm to calculate the Shapley values

Input: Number of iterations M, instance of interest x, feature index j, data matrix X, and dual input CNN model

for *Every Iteration 1 ..... M* **do**

Draw random instance z from the data matrix X

Choose a random permutation 0 of the feature values

Order instance x: $(x_0 ...., x_j, ... , x_p)$

Order instance z: $(z_0 ...., z_j, ... , z_p)$

Construct two new instances:

with j : $x_{-j} = (x_0..., x_j, z_{j+1}, ...z_p)$

without j : $x_{+j} = (x_0..., x_j, z_{j+1}, ...z_p)$

Compute marginal distribution:

$$\phi_j^m = \hat{f}(x_{+j}) - \hat{f}(x_{-j}) \tag{6.1}$$

end for

Compute Shapley values:

$$\phi_j(x) = \frac{1}{M} \sum_{m=1}^{M} \phi_j^m \tag{6.2}$$

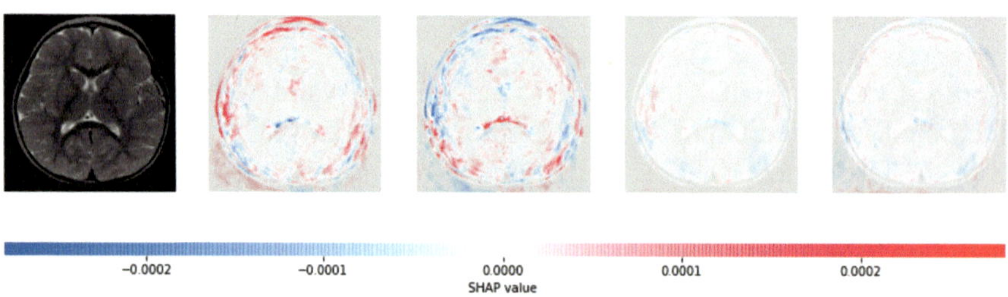

Figure 6.4 On the basis of Shapley values, we can say that the MRI image is normal.

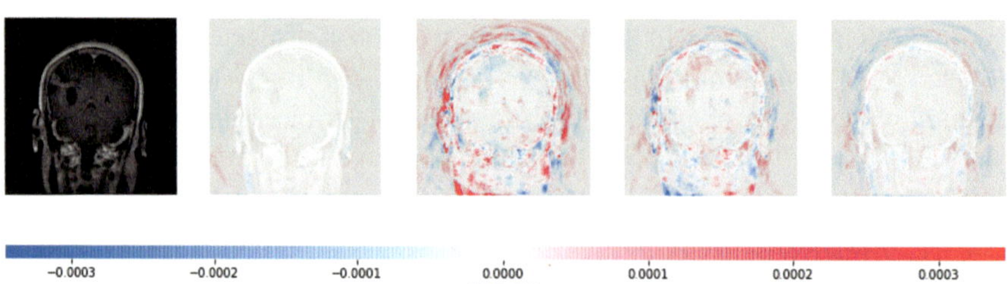

Figure 6.5 On the basis of Shapley values, we can say that the MRI image hold meningioma tumor.

the likelihood of the class. The first two figures, Figures 6.4 and 6.5, are test images. In contrast, the remaining figures show the normal image and three more tumor categories: glioma, meningioma, and pituitary tumors, in successive order.

6.3.3 LIME

We applied 150 alterations overall. Superpixels act as columns and perturbations as rows in a matrix created from a random set of ones and zeros. Superpixels are ON if they are one and OFF if they are zero. The number of superpixels in the image is indicated by the length of the displayed vector. Based on the perturbation vector and predefined superpixels, the test image is altered [35]. The final perturbed image is displayed in Figure 6.6(C) for a normal test image under consideration and in Figure 6.7(C) for a test image

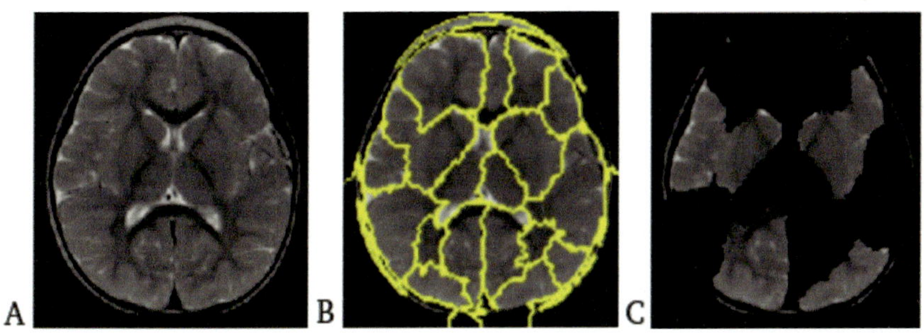

Figure 6.6 Interpretations generated by LIME for normal image. (A) Sample of the normal image from the test image. (B) Superpixels generated from sample of the normal image from test image quick-shift segmentation to create perturbations. (C) Final perturbed image for normal image.

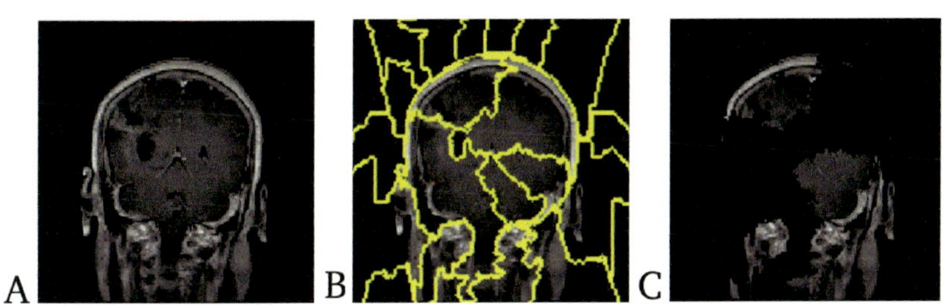

Figure 6.7 Interpretations generated by LIME for meningioma tumor. (A) Sample of meningioma tumor from the test image. (B) Superpixels generated from quick-shift segmentation to create perturbations. (C) Final perturbed image showing meningioma tumor.

under consideration with meningioma tumor, which demonstrates the area of the image that plays a key role in classification.

For Figures 6.6–6.7, the explanation is produced using LIME by the CNN model. Figure 6.6(A) shows the sample of the normal image from the test image and Figure 6.7(A) shows the sample of meningioma tumor from the test image. For each category that is accessible in the CNN model, the classification yields a vector of 2870 probabilities. Superpixels are produced using the quick-shift segmentation technique. The 22 superpixels are generated for Figure 6.6(A) and displayed in Figure 6.6(B) and 24 superpixels are computed for Figure 6.7(A) and shown in Figure 6.7(B).

6.4 Discussion

6.4.1 Comparison of proposed feature extraction methods using traditional machine learning (ML) methods

We contrast conventional ML algorithms with our proposed feature extraction techniques. The results of the comparison are shown in Table 6.3. The authors in [24] used GLCM (gray-level co-occurrence matrix) for feature extraction and classification boosting (Adaboost) after preprocessing the MICCAI BRaTs dataset to remove noise. Twenty-two traits were extracted using an MRI. The suggested method has a maximum accuracy of 89.90% and uses the Adaboost classifier for classification. After gathering the MRI images, the authors in [1] used a computer-aided detection system. Prior to applying the back-propagation algorithm, they analyzed the image and used Haralick's features based on the spatial gray-level dependency matrix (SGLD) to extract the features. Results were 99% accurate, although the study was unable to focus on the explainable component of the training images. The classification of benign and malignant tumors was done using a comparison of random forest (RF) and support vector machine (SVM). First, identifying the area of interest for feature extraction from the brain tumor, then extracting features. To differentiate between benign and malignant tumors, shape characteristics have been obtained and put to use. The authors found that RF (81.90%) performed better than the SVM (78.57%). Using the discrete wavelet transform (DWT), principal component analysis (PCA), KSVM, and GRB kernels, the authors of [2] were used to provide a novel method for categorizing MRI brain pictures. Four different kernels were used for the experiment. The results show that, in comparison to other approaches, the combination of DWT, PCA, KSVM, and the GRB kernel produces the maximum accuracy. The findings indicate a significant reduction in classification time for segmented images, which could mark a turning point for the area of tumor diagnosis in medical field. The BRaTs 2015 training dataset was used by the authors in [22] to work on the FLAIR pictures; it is used to reorganize and enhance the data attributes leading to a pixel-based classifier. In comparison to the class balancing approach used by the random forest classifier with synthetic minority oversampling technique (SMOTE), the U-net proposed method achieves semantic segmentation with a precision of 76%, an increase of 23%.

Table 6.3 Brain tumor detection using traditional ML methods.

Authors	Algorithm	Dataset	Accuracy (%)	XAI
Martinez et al. [22]	Random forest	BRaTs dataset	76	No
Minz and Mahobiya [24]	Adaboost classifier	BRaTs dataset	89.90	No
Abdalla and Esmail [1]	Back-propagation network	MRI images	**99**	**No**
Asodekar and Gore [3]	Random forest	BRaTs dataset	81.90	No
Asodekar and Gore [3]	SVM	BRaTs dataset	78.57	No
Proposed model	**Dual Input CNN**	MRI images	**94.64**	**Yes**

6.4.2 Comparison of the proposed method with the other state-of-the-art methods

In this section, our dual input CNN model is compared to various cutting-edge models. Table 6.4 compares the outcomes. The DL model was put into practice by researchers [16] after a number of data collecting and preprocessing procedures, such as segmentation and average filtering. The research represents overall performance and comparative output on the brain MRI images when compared to existing methodologies like conditional random field (89%), SVM (84.5%), and genetic algorithm (GA) (83.64%). The proposed CNN (91%) outperforms existing algorithms in terms of results. A DL technique termed faster R-CNN in [4] was built using the TensorFlow framework, and the classifier algorithm was trained and evaluated using a publically available dataset of 3064 MRI brain images (708 meningiomas, 1426 gliomas, and 930 pituitary gland) from 233 patients. The faster R-CNN method has been shown to achieve an accuracy of 91.66%, which is superior to the earlier work on the same dataset. A cascaded convolutional neural network (C-ConvNet/C-CNN) was suggested by the authors in [26]. The CNN model has been proposed to extract local and global characteristics in two techniques, with distinct extraction patches in each. It is a straightforward but efficient cascade. A complex preprocessing method was used to extract the expected position of tumor, and those patches were selected to alert the network that the center of tumor was situated in this region. The computing time and ability to generate quick predictions for categorizing the clinical image is reduced due to the removal of a large number of irrelevant pixels from the image during preprocessing. The output of other algorithms was compared to the results. Even yet, the CNN model had the highest overall Dice score (mean) accuracy of 92.03% and the highest core Sensitivity score (mean) precision of 97.12%.

One study [19] uses a hybrid model of CNN and SVM for classification, type, and threshold-based segmentation in terms of detection to distinguish between benign and malignant tumors in brain MRI images. The overall accuracy of this hybrid CNN-SVM is scored at 98.49%. However, their work does not provide any proof that XAI and low-quality images can be altered. The classification of 3D brain tumor MRIs into HG and LG gliomas was suggested by the authors in [30] using a CNN cascade with a long short-term memory (LSTM) network. An LSTM network was fed with the features from the pre-trained VGG-16 in order to learn high-level feature representations. In comparison to the 71% classification accuracy of the components recovered from AlexNet and ResNet, the VGG-16 components had an accuracy of 84%. In this article [17], using a 1616 PatchGAN, the authors looked into conditional adversarial networks as a general-purpose approach to the problem of image-to-image translation. These distortions are lessened and scores are marginally improved by the PatchGAN 70×70. It has been noted that scaling the results to the full 286×286 ImageGAN does not significantly enhance their visual impact. The significantly reduced FCN-score illustrates the potential of conditional adversarial networks for numerous image-to-image translation tasks, especially those requiring complex structured graphical results.

The authors in [23] presented a method for segmenting 3D images that rely on a volumetric, fully convolutional neural network. CNN predicts segmentation for the entire volume at once after being trained end-to-end on MRI volumes indicating prostate. The PROMISE2012 challenge dataset served as the source for the relative manual ground truth annotation, which was trained on 50 MRI volumes. The novel objective function was to maximize training results based on the dice overlap coefficient comparing the predicted segmentation and the ground truth annotation. An unsupervised medical anomaly detection generative adversarial network (MADGAN) was proposed by the authors of [15]. To identify brain anomalies on multi-sequence structural MRI at different phases, their two-step method leverages GAN-based multiple neighboring brain MRI slice reconstruction. With an area under the curve (AUC) of 0.727 for mild cognitive impairment (MCI), 0.894 for late-stage AD, and 0.921 for brain metastases on T1c scans, MADGAN can detect anomalies on T1 scans at an incredibly early stage. As the first unsupervised unique disease diagnosis, the model may successfully identify the accumulation of minor anatomical abnormalities and hyper-intense enhancing lesions, such as (especially late-stage) AD and brain metastases, using multi-sequence MRI. By incorporating a model of healthy anatomy into a CycleGAN-based style-transfer task, which is trained to convert healthy brain MRI images into a simulated distribution with reduced entropy and vice versa, the authors of [6] provided a unique approach to unsupervised AD in brain MRI. The resulting model reduces anomalies when reviving the input data at test time by filtering high-frequency, low-amplitude signals from lower entropy samples during training. The approach overcomes current limitations of autoencoder (AE)-based approaches and outperforms the state-of-the-art in a variety of metrics when dealing with high-resolution data. The authors in [8] focus on the problems that need to be solved in order to develop AI applications like clinical decision support systems in an operational setting. A narrative analysis and critical evaluation of publications released between 1989 and 2021 were done. The study found that biomedical and healthcare systems were among the most important categories for AI applications, with medical imaging being the best-suited and most promising domain. The development of such systems and their application to clinical practice are made easier by the clarification of particular challenging issues. The authors in [5] presented the foundational technologies of AI as well as state-of-the-art approaches and how they were applied to medical imaging. This article provided an overview of AI emphasizing its application in medical imaging analysis. It highlighted how state-of-the-art ML and DL algorithms possess the capability to automate and enhance various tasks related to clinical practice.

Table 6.4 Brain tumor detection using other state-of-the-arts.

Authors	Algorithm	Dataset	Accuracy (%)	XAI
Shahzadi et al. [30]	CNN with LSTM	MRI images	84	No
Hemanth et al. [16]	CNN	MRI images	91	No
Avsar and Salcin [4]	R-CNN	MRI images	91.66	No
Ranjbarzadeh et al. [26]	C-CNN	BRaTs dataset	92.03	No
Khairandish et al. [19]	CNN-SVM	MRI images	**98.49**	**No**
Proposed model	**Dual input CNN**	MRI images	**94.69**	**Yes**

6.5 Conclusion and future direction

The proposed study attained an accuracy of 94.64% when determining whether a specific MRI image indicates a tumor or not using an explanation-driven dual input CNN model. The proposed CNN model is trained and tested using a dataset of brain MRI images, and the same model is also applied to the SHAP and LIME methods to provide an explanation. For better feature extraction in our experiment, we used two dataset copies as input — one in the convolution layer and another in the fully connected layer. The overall performance of the prediction model was negatively affected by any attempt to eliminate any features; hence, no augmentation was made. The proposed framework is a locally interpreted model with a model-agnostic explanation which is clearly stated to more qualitatively depict the outcomes for typical individuals.

As a future work, XAI can be enabled to incorporate classification algorithms with higher accuracy and better optimizer. The research could be repeated and applied to other XAI algorithms, such as GradCAM, to improve clinical problems. Furthermore, algorithms that simulate natural occurrences can be applied to heterogeneous datasets for medical imaging modalities, electronic health record engines, multi-omics studies, and real-time monitoring [29], similar to most recent developments in computational capability, neuroimaging technologies, and digital phenotyping tools [27].

References

[1] Hussna Elnoor Mohammed Abdalla and Mohammed Yagoub Esmail. Brain tumor detection by using artificial neural network. pages 1–6, 2018.

[2] Preeti Arora and Rajeev Ratan. Development of a novel approach for classification of mri brain images using dwt by integrating pca, ksvm and grb kernel. 2021.

[3] Bhagyashri Asodekar and Sonal A. Gore. Brain tumor classification using shape analysis of mri images. Jul 2019.

[4] Ercan Avsar and Kerem Salcin. Detection and classification of brain tumours from mri images using faster r-cnn. volume 13, pages 337–342, 12 2019.

[5] Ana Barragán-Montero, Umair Javaid, Gilmer Valdés, Dan Nguyen, Paul Desbordes, Benoit Macq, Siri Willems, Liesbeth Vandewinckele, Mats Holmström, Fredrik Löfman, Steven Michiels, Kevin Souris, Edmond Sterpin, and John A. Lee. Artificial intelligence and machine

learning for medical imaging: A technology review. *Physica medica: PM: an international journal devoted to the applications of physics to medicine and biology: official journal of the Italian Association of Biomedical Physics (AIFB)*, 83:242–256, March 2021.

[6] Christoph Baur, Robert Graf, Benedikt Wiestler, Shadi Albarqouni, and Nassir Navab. Steganomaly: Inhibiting cyclegan steganography for unsupervised anomaly detection in brain mri. In Anne L. Martel, Purang Abolmaesumi, Danail Stoyanov, Diana Mateus, Maria A. Zuluaga, S. Kevin Zhou, Daniel Racoceanu, and Leo Joskowicz, editors, *Medical Image Computing and Computer Assisted Intervention – MICCAI 2020*, pages 718–727, Cham, 2020. Springer International Publishing.

[7] Sartaj Bhuvaji. Brain tumor classification-mri. volume 2, may 2020.

[8] Isabella Castiglioni, Leonardo Rundo, Marina Codari, Giovanni Di Leo, Christian Salvatore, Matteo Interlenghi, Francesca Gallivanone, Andrea Cozzi, Natascha Claudia D'Amico, and Francesco Sardanelli. Ai applications to medical images: From machine learning to deep learning. *Physica Medica*, 83:9–24, 2021.

[9] Devam Dave, Het Naik, Smiti Singhal, and Pankesh Patel. Explainable AI meets healthcare: A study on heart disease dataset. volume abs/2011.03195, 2020.

[10] Gaillard Frank. Glioblastoma: Radiology reference article. 2021.

[11] Frank Gaillard. Normal brain mri: Radiology case. 2021.

[12] Frank Gaillard and Bahman Rasuli. Meningioma. 2021.

[13] Razdan T Mallik S Zhao Z Gaur L, Bhandari M. Explanation-driven deep learning model for prediction of brain tumour status using mri image data. *Frontier in Genetics*, 2022.

[14] Manu Gupta, Prabhakar Rao, and Venkateswaran Rajagopalan. Brain tumor detection in conventional mr images based on statistical texture and morphological features. In *2016 International Conference on Information Technology (ICIT)*, pages 129–133, 2016.

[15] Changhee Han, Leonardo Rundo, Kohei Murao, Tomoyuki Noguchi, Yuki Shimahara, Zoltan Milacski, Saori Koshino, Evis Sala, Hideki Nakayama, and Shinichi Satoh. Madgan: unsupervised medical anomaly detection gan using multiple adjacent brain mri slice reconstruction. *BMC Bioinformatics*, In Press, 11 2020.

[16] G. Hemanth, M. Janardhan, and L. Sujihelen. Design and implementing brain tumor detection using machine learning approach. In *2019 3rd International Conference on Trends in Electronics and Informatics (ICOEI)*, pages 1289–1294, 2019.

[17] Phillip Isola, Jun-Yan Zhu, Tinghui Zhou, and Alexei A. Efros. Image-to-image translation with conditional adversarial networks. 2018.

[18] Qiao Ke, Jiangshe Zhang, Wei Wei, Robertas Damasevicius, and Marcin Woźniak. Adaptive independent subspace analysis of brain magnetic resonance imaging data. *IEEE Access*, PP:12252–12261, 02 2019.

[19] M.O. Khairandish, M. Sharma, V. Jain, J.M. Chatterjee, and N.Z. Jhanjhi. A hybrid cnn-svm threshold segmentation approach for tumor detection and classification of mri brain images. 2021.

[20] Diederik P. Kingma and Jimmy Ba. Adam: A method for stochastic optimization. In Yoshua Bengio and Yann LeCun, editors, *3rd International Conference on Learning Representations, ICLR 2015, San Diego, CA, USA, May 7-9, 2015, Conference Track Proceedings*, 2015.

[21] Dimitrios Kollias, Athanasios Tagaris, Andreas Stafylopatis, Stefanos Kollias, and Georgios Tagaris. Deep neural architectures for prediction in healthcare. volume 4, pages 119–131, Jun 2018.

[22] Elizabeth Martinez, Camilo Calderón, Hans Garcia, and Henry Arguello. Mri brain tumour segmentation using a cnn over a multi-parametric feature extraction. In *2020 IEEE Colombian Conference on Applications of Computational Intelligence (IEEE ColCACI 2020)*, pages 1–6, 2020.

[23] Fausto Milletari, Nassir Navab, and Seyed-Ahmad Ahmadi. V-net: Fully convolutional neural networks for volumetric medical image segmentation. 2016.

[24] Astina Minz and Chandrakant Mahobiya. Mr image classification using adaboost for brain tumor type. In *2017 IEEE 7th International Advance Computing Conference (IACC)*, pages 701–705, 2017.

[25] Arvind Neelakantan, Luke Vilnis, Quoc V. Le, Ilya Sutskever, Lukasz Kaiser, Karol Kurach, and James Martens. Adding gradient noise improves learning for very deep networks, 2015.

[26] Ramin Ranjbarzadeh, Abbas Bagherian Kasgari, Saeid Jafarzadeh Ghoushchi, Shokofeh Anari, Maryam Naseri, and Malika Bendechache. Brain tumor segmentation based on deep learning and an attention mechanism using MRI multi-modalities brain images. volume 11. Springer Science and Business Media LLC, 2021.

[27] Kerry J. Ressler and Leanne M. Williams. Big data in psychiatry: Multi-omics, neuroimaging, computational modeling, and digital phenotyping, Sep 2020.

[28] Antoine Richard, Brice Mayag, François Talbot, Alexis Tsoukias, and Yves Meinard. Transparency of classification systems for clinical decision support. volume 1239, pages 99–113, 2020.

[29] Leonardo Rundo, Carmelo Militello, Salvatore Vitabile, Giorgio Russo, Evis Sala, and Maria Gilardi. A survey on nature-inspired medical image analysis: A step further in biomedical data integration. *Fundamenta Informaticae*, 171:345–365, 10 2019.

[30] Iram Shahzadi, Tong Boon Tang, Fabrice Meriadeau, and Abdul Quyyum. Cnn-lstm: Cascaded framework for brain tumour classification. In *2018 IEEE-EMBS Conference on Biomedical Engineering and Sciences (IECBES)*, pages 633–637, 2018.

[31] Yukti Sharma, Abhinav Verma, Krisstina Rao, Vivek Eluri, Abhinav Verma, Krisstina Rao, Vivek Eluri, and Yukti Sharma. 'reasonable explainability' for regulating ai in health. Oct 2020.

[32] Pembury Smith, Q. R. Matilda, and Graeme D. Ruxton. Effective use of the mcnemar test. *Behavioral Ecology and Sociobiology*, 74, 10 2020.

[33] Angela Spatharou, Solveigh Hieronimus, and Jonathan Jenkins. Transforming healthcare with ai: The impact on the workforce and organizations. 03 2021.

[34] Hyuna Sung, Jacques Ferlay, Rebecca L. Siegel, Mathieu Laversanne, Isabelle Soerjomataram, Ahmedin Jemal DMV, and Freddie Bray. Global cancer statistics 2020. 2021.

[35] Andrea Vedaldi and Stefano Soatto. Quick shift and kernel methods for mode seeking. In David Forsyth, Philip Torr, and Andrew Zisserman, editors, *Computer Vision – ECCV 2008*, pages 705–718, Berlin, Heidelberg, 2008. Springer Berlin Heidelberg.

[36] Yuranga Weerakkody and Frank Gaillard. Pituitary adenoma: Radiology reference article. 2021.

[37] Samek Wojciech, Wiegand Thomas, and Müller Klaus Robert. Explainable artificial intelligence: Understanding, visualizing and interpreting deep learning models. 2017.

[38] Sun Yu, Lin Zhu, Guan Wang, and Fang Zhao. Multi-input convolutional neural network for flower grading. volume 2017, page 8, 2017.

[39] Chiara Zucco, Huizhi Liang, Giuseppe Di Fatta, and Mario Cannataro. Explainable sentiment analysis with applications in medicine. In *2018 IEEE International Conference on Bioinformatics and Biomedicine (BIBM)*, pages 1740–1747, 2018.

7

Conclusions and Scope for Further Research

Conclusions

In this book, we have discussed how pattern recognition methods are used in next-generation sequencing and medical imaging data. In this book, we have discussed, fundamental things of next-generation sequencing, pattern recognition, and biomedical imaging in our introductory chapter (i.e., Chapter 1), and in the remaining chapters (i.e., Chapter 2–6), we have discussed some recently developed research works on next-generation sequencing and medical imaging data using the tools of pattern recognition and machine learning. Pattern recognition has so many applications in various real-world scenarios, significantly enhancing automation and decision-making processes. As example, multi-objective optimized variable cutoff based association rule mining image and object recognition, speech and audio processing, natural language processing, medical diagnosis, financial analysis, and biometrics are some areas where pattern recognition tools are very essential. The important phases of pattern recognition which people used in frameworks of research development are data acquisition, feature extraction, pre-processing, classification/clustering, training and learning, validation and testing, postprocessing, and decision-making. In Chapter 1, we have introduced next-generation sequencing data (NGS), basic techniques of pattern recognition, and the tools of biomedical imaging.

In Chapter 2, we have presented an integrated approach of linear regression and deep learning for detecting reliable genetic alterations in cancer using DNA methylation and gene expression data. Finding cancer biomarkers, classifying them, and developing new treatments have all benefited from changes in DNA methylation. The approaches currently in use either use combined CpG sites or differentially methylated regions, which can be mapped to genes, that are composed of CpG sites that have undergone various levels of methylation. Such methylation signal mapping does have certain drawbacks. To overcome this, we developed a combinatorial framework for accurate biological interpretation of DNA methylation through integrating DNA methylation data and related TCGA gene expression data. In this framework, we used linear regression, differential expression, and deep learning methods to solve these constraints. We used cervical uterine cancer to demonstrate it. We pre-filtered outliers from the dataset before using linear regression to extract the estimated gene exprcssion value from the pre-filtered methylation data. Using the empirical Bayes test and *Limma*, we discovered differentially expressed genes (DEGs). Then, using 10-fold crossvalidation, we employed the deep learning technique *nnet* to identify the cervical cancer labels of DEGs in order to calculate all classification metrics, including accuracy and area under the curve (AUC). Here we used DNA methylation data from uterine cervical cancer (NCBI Accession ID: GSE30760; 27,578 features covering 63 tumor and 152 matched normal samples). Following differential expression analysis and linear regression, we discovered 6287 DEGs with a false discovery rate (FDR) < 0.001. After doing deep learning analysis, we discovered that the uterine cervical malignant labels had an average classification accuracy of $90.69\%(\pm 1.97\%)$. When compared to other peer approaches, its performance is superior. Using *Cytoscape*, we conducted in-degree and out-degree hub gene network analysis. We identified the top five in-degree (*PAIP2*, *GRWD1*, *VPS4B*, *CRADD*, and *LLPH*) and top five out-degree (*MRPL35*, *FAM177A1*, *STAT4*, *ASPSCR1*, and *FABP7*) genes. After that, we used the tool $WebGestalt(Web - basedGeneSeTAnaLysisToolkit)$ to perform KEGG pathway and gene ontology enrichment analysis of DEGs. It is proven that, this framework offers an effective strategy for better interpreting data from gene expression and DNA methylation analyses in disease studies.

Epigenetic signatures may be undirectional (up-regulated or down-regulated) and it can also be directional. In this book, Chapter 3 explains an integral strategy for the detection of directional signatures using association rule mining technique and ranking them through multi-objective optimization. A strong approach for studying the informative relationships between different features (genes) in any dataset is association rule mining (ARM). The fact that ARM creates a large number of rules with various rule-informative values makes it difficult for the user to select the most useful rules. A limited number of studies have also been conducted on the integration of various biological datasets and adaptable cutoff values in ARM. We stated a novel framework $MOOVARM$ (multi-objective optimized variable cutoff-based association rule mining) for multi-omics profiles to address each of these issues. For all genesets (itemsets) associated with each extracted rule, this framework discovered the positive ideal solution (PIS) that maximized the profit and minimized the loss as well as the negative ideal solution (NIS) that did exactly the reverse. Then, for each geneset or product, we calculated the distance ($d+$) from PIS and the distance ($d-$) from NIS. In the multi-omics dataset, these two distances were crucial in determining the most optimal associations between various pairings of genes. Then, in order to rank the genesets, we evaluated the relative closeness to the PIS globally. The rule can be regarded as a final resultant rule if its relative closeness score is greater than or equal to the predetermined threshold value. Additionally, rather than solely looking at individual genes, $MOOVARM$ evaluated the relative score of the rule depending on the status of every gene. Finally, $MOOVARM$ generated the final rank of the extracted (multi-objective optimized) rules of associated genes, which showed superior disease classification when compared to the most robust gene signature identification algorithms.

Nowadays, sequencing has been changed from normal RNA sequencing to single-cell RNA sequencing since the maximum data is from heterogeneous population. The gene expression profiles of individual cells originating from homogeneous and heterogeneous populations are analyzed by single-cell RNA sequencing, which can isolate single cells and sequence the RNA from each cell separately. Additionally, cellular heterogeneity and rare cell populations that might be hidden in normal RNA-seq data can be revealed by single-cell RNA sequencing. Identification of existing and novel cell types, cell representation, cell fate prediction, classification of various tumor types, and study of heterogeneity in various cells are the key areas of interest for single-cell RNA sequential analysis. Due to the nature of the data, cluster identification in single-cell sequencing data with high dimensions presents several difficulties. Techniques of dimensionality reduction can address the issue. In Chapter 4, we have discussed a framework for potential cluster specified frequent biomarkers discovery using dimensionality reduction and hierarchical agglomerative clustering Louvain for single-cell RNA sequencing data analysis. First, we performed a pre-filtering operation to remove features with less cells and cells with less features. Then, using quality control measures, we discarded cells that were of low quality or dying and generated a Seurat object to store the data and analysis together. After that, we normalized the data using the global-scaling normalization method "LogNormalize." Next, using data from our dataset, we generated highly variable features from cell to cell. Then, to project high-dimensional data into an ideal low-dimensional space, we used a linear transformation and linear dimensionality reduction approach called principal component analysis (PCA). We used a graph-based clustering approach called Louvain for the cell clustering of the top 10 significant principal components (PCs) from our already detected 50 "significant" PCs based on notable enrichment of low p-value features. A single-cell RNA sequencing dataset for a rare type of intestinal cell in mice (NCBI Accession ID: GSE62270, 23,630 features, and 1872 samples (cells)) was used to test this approach. Ten cell clusters with a maximum modularity of 0.8851 were produced. Using model-based analysis of single-cell transcriptomics (MAST), an expression feature extraction statistical tool for single-cell sequencing data, with a log_2FC threshold of 0.25 and a minimum feature detection of 25%, we found 3871 cluster-specific biomarkers after identifying the cell clusters. We identified 1892 most frequent markers – overlapping biomarkers – from these cluster-specific biomarkers. Using Cytoscape, we conducted a degree hub gene network analysis and identified the five highest degree genes: $Rps4x$, $Rps18$, $Rpl13a$, $Rps12$, and $Rpl18a$. Using the David 6.8 software tool, we then carried out KEGG pathway and gene ontology enrichment analyses on cluster markers. In conclusion, our framework, which combines dimensionality reduction with agglomerative hierarchical clustering, offers a reliable method for quickly identifying overlapping biomarkers, or cluster-specific frequent biomarkers, from single-cell RNA sequencing data.

Three basic functions, including the processing of genetic information, the formation of cell membranes, and the production of energy, gene translation, deoxyribonucleic acid replication, and fundamental cellular structure and facilitate intracellular and extracellular transport are carried out by essential genes in single-celled organisms. For different species to survive, essential genes are crucial. These genes belong to a family that is connected to essential cellular functions for species survival. Significant genomics data is preserved by essential genes, which may hold the key to a thorough understanding of life and evolution. Due to its importance, fundamental gene studies have long been regarded as a crucial area in computational biology. Every possible combination of adenine, guanine, cytosine, and thymine builds an essential gene. In this book, Chapter 5 introduces an approach for identifying the stationary patterns of nucleotides like adenine, guanine, cytosine, and thymine in each gene. The statistical distribution of stationary patterns of nucleotides in the genes is brought to these purposes in a set of co-occurrence matrices, which are useful in determining the relationship between the nucleotides. Energy, entropy, homogeneity, contrast, and dissimilarity features are computed for the purpose of capturing discriminant traits from each co-occurrence matrix. These features are then retrieved from all co-occurrence matrices and concatenated to produce a feature vector representing each essential gene. On the basis of the extracted fixed dimensional feature vectors, supervised machine learning methods have finally been employed for essential gene classification. Some current state-of-the-art feature encoding techniques, including Shannon entropy, Hurst exponent, fractal dimension, and their combinations, have been used for comparative study. The robustness and efficiency of the developed methods have been demonstrated through a thorough experiment that classified the essential genes of five species.

Deep learning (DL) techniques for assessing magnetic resonance imaging (MRI) images and predicting brain cancers have experienced accelerated growth in cancer research. In Chapter 6, we have presented a deep learning model for prediction of brain tumor status using MRI image data, since for DL models, we have seen a substantial disparity in explanation, interpretability, and high accuracy. For the prediction of unique subtypes of brain tumors (meningioma, glioma, and pituitary), we used a convolutional neural network (CNN), local interpretable model-agnostic explanation (LIME), and Shapley additive explanation (SHAP) in our explanation-driven DL model. Our model, in contrast to earlier models, added Gaussian noise to images of lower quality in terms of noise and metal artifacts in order to overcome the classification challenge. Comparing our CNN training results to other state-of-the-art, they show 94.64% accuracy. By examining all future forecasts using any possible combination of inputs, we applied SHAP to assure consistency and local accuracy for interpretation. In contrast, LIME builds sparse linear models around each prediction to show how the model performs locally. In this study, interpretability and high accuracy are prioritized because they are crucial for identifying inequalities in prediction performance, advantageous for fostering trust, and necessary for integration into clinical practice. This approach has a broad therapeutic application that might be utilized for mass screening in nations with limited resources.

Future Research

In order to create a better classification model across all domains, as a future work, we can develop a deep learning technique that incorporates the random forest ensemble method and apply that classifier with big data or unstructured data for cancer classification. We are also interested in extending the rule mining framework by including fuzzy/rough theory and multi-objective optimization together to produce better and more effective association rule mining from multi-omic data as well as big data. Since the drop-out problem is a critical issue of single-cell sequencing data, we want to improve the framework by importing data imputation and multi-objective optimization in clustering. Besides biomarkers detection, we are interested in applying our framework for rare cell detection. We are also interested in updating the XAI framework with higher accuracy and better optimizer-based classifier. Deep learning and random forest ensemble classifier may be a better choice for this purpose. Besides, the most recent advances in computing capacity, neuroimaging technologies, and digital phenotyping tools enable the simulation of natural events. These advancements facilitate work on heterogeneous datasets including those from medical imaging modalities, electronic health records engines, multi-omics studies, and real-time monitoring.

Index

About the Authors

Dr. Saurav Mallik is currently working as Research Scientist in the Department of Pharmacology and Toxicology, The University of Arizona, USA. Previously, he worked as Postdoctoral Fellow in Harvard T.H. Chan School of Public Health, University of Texas Health Science Center at Houston, and University of Miami Miller School of Medicine, USA. He obtained his Ph.D. degree in the Department of Computer Science and Engineering from Jadavpur University, Kolkata, India in 2017 with his Ph.D. works carried out in the Machine Intelligence Unit, Indian Statistical Institute, Kolkata, India as Junior Research Fellow and Visiting Scientist. He obtained the award of Research Associateship from CSIR (Council of Scientific and Industrial Research), MHRD, Govt. of India in 2017. Dr. Mallik has published more than 130 research papers in different top high impact factor peer-reviewed international journals, conferences and book chapters having H-index 19. He is working as the active member of Institute of Electrical and Electronics Engineers (IEEE), USA, ACM and American Association for Cancer Research (AACR), USA and Bioclues, India. He has also worked with section editors and reviewers with several well-reputed high impact journals. His research interest includes computational biology, knowledge retrieval and data mining, bioinformatics, bio-statistics and machine learning/deep learning.

Prof. Loveleen Gaur is currently working as Professor, Program Director Artificial Intelligence and Data Analytics at Amity University, India. She has more than 20 years of teaching, research, and administrative experience internationally. She is the founding director of the postgraduate programme in Artificial Intelligence and Data Analytics in Amity International Business School. She is supervising a number of Ph.D. scholars, Post Graduate students, mainly in Artificial Intelligence and Data Analytics for business and management. Under her guidance, the AI/Data Analytics research cluster has published extensively in high-impact factor journals and has established extensive research collaboration globally with several renowned professionals. She is a senior IEEE member and Series Editor with CRC and Wiley. She has high indexed publications in SCI/ABDC/WoS/Scopus, has several patents/copyrights to her credit, and has edited/authored more than 20 research books published by world-class publishers. An ample number of Ph.D. and Master's students have graduated under her supervision. She is an external Ph.D./Master thesis examiner/evaluator for several universities globally. She has completed internationally funded research grants successfully. Professor Gaur has significantly contributed to enhancing scientific understanding by participating in over three hundred scientific conferences, symposia, and seminars, by chairing technical sessions and delivering plenary and invited talks.

Dr. Soumita Seth is currently serving as an Assistant Professor in the Department of Computer Science and Engineering of Future Institute of Engineering and Management, Kolkata, India, Affiliated to MAKAUT, Kolkata. Previously, she worked as Assistant Professor in the Department of Computer Science and Engineering of Brainware University, Kolkata (March 2023 to August 2023) and in the Department of Computer Science and Engineering of Pailan College of Management and Technology, Kolkata, India, Affiliated to MAKAUT, Kolkata (February 2022 to March 2023). In addition, she recently submitted her Ph.D. thesis to the Department of Computer Science & Engineering (CSE) from a state-government university, Aliah University (AU), Kolkata, India. Previously, she completed M.Tech. and B.Tech. from the departments of CSE and IT, respectively. She is also collaborating her Ph.D. research with The University of Texas Health Science Center at Houston (UTHealth), Houston, USA. She has academic experience of almost 6 years, research experience of 2 years, and industrial experience of 2 years. Dr. Seth has more than 10 research papers in different top high impact factor peer-reviewed international journals, conferences and book chapters. She has also worked with section editors and section reviewers with several well-reputed high impact journals. Her research interests include computational biology, data mining, bioinformatics, pattern recognition, biological regulatory networks, statistical application on bioinformatics, machine learning/deep learning.

Dr. Tapas Bhadra has worked as an Assistant Professor in the Department of Computer Science and Engineering, Aliah University, Kolkata, India since 2015. He completed Ph.D. from the Department of Computer Science and Engineering of Jadavpur University, Kolkata, India in 2018. Prior to this, he was an INSPIRE Fellow in the Machine Intelligence Unit, Indian Statistical Institute, Kolkata from 2011 to 2015. He has co-authored more than 30 research papers having H-index 7. His research interests include pattern recognition, data mining, computational biology, and bioinformatics.

Dr. Mingqiang Wang is currently working as postdoctoral scholar in the Department of Cardiovascular Institute, School of Medicine, Stanford University, USA. He previously worked in the Department of Center of Precision Health in University of Texas Health Science Center, Houston, Texas, USA as Postdoctoral Research Fellow. He obtained his Ph.D. degree in the Department of Bioinformatics from Chinese University of Hong Kong, Shatin, Hong Kong in 2018 and Master in Science from Chinese Academy of Sciences, Beijing, China in 2016. Dr. Wang has published many articles in various top international journals and conferences associated with bioinformatics. His research interest includes bioinformatics and computational genomics.

For Product Safety Concerns and Information please contact our EU representative GPSR@taylorandfrancis.com Taylor & Francis Verlag GmbH, Kaufingerstraße 24, 80331 München, Germany